Principles Of

Environmental Analysis

HARVESTIME INTERNATIONAL INSTITUTE

This course is part of the **Harvestime International Institute**, a program designed to equip believers for effective spiritual harvest.

The basic theme of the training is to teach what Jesus taught, that which took men who were fishermen, tax collectors, etc., and changed them into reproductive Christians who reached their world with the Gospel in a demonstration of power.

This manual is a single course in one of several modules of curriculum which moves believers from visualizing through deputizing, multiplying, organizing, and mobilizing to achieve the goal of evangelizing.

TABLE OF CONTENTS

HOW TO USE THIS MANUAL

MANUAL FORMAT

Each lesson consists of:

Objectives: These are the goals you should achieve by studying the chapter. Read them before starting the lesson.

Key Verse: This verse emphasizes the main concept of the chapter. Memorize it.

Chapter Content: Study each section. Use your Bible to look up any references not printed in the manual.

Self-Test: Take this test after you finish studying the chapter. Try to answer the questions without using your Bible or this manual. When you have concluded the Self-Test, check your answers in the answer section provided at the end of the book.

For Further Study: This section will help you continue your study of the Word of God, improve your study skills, and apply what you have learned to your life and ministry.

Final Examination: If you are enrolled in this course for credit, you received a final examination along with this course. Upon conclusion of this course, you should complete this examination and return it for grading as instructed.

ADDITIONAL MATERIALS NEEDED

You will need a King James version of the Bible.

SUGGESTIONS FOR GROUP STUDY

FIRST MEETING

Opening: Open with prayer and introductions. Get acquainted and register the students.

Establish Group Procedures: Determine who will lead the meetings, the time, place, and dates for the sessions.

Praise And Worship: Invite the presence of the Holy Spirit into your training session.

Distribute Manuals To Students: Introduce the manual title, format, and course objectives provided in the first few pages of the manual.

Make The First Assignment: Students will read the chapters assigned and take the Self-Tests prior to the next meeting. The number of chapters you cover per meeting will depend on chapter length, content, and the abilities of your group.

SECOND AND FOLLOWING MEETINGS

Opening: Pray. Welcome and register any new students and give them a manual. Take attendance. Have a time of praise and worship.

Review: Present a brief summary of what you studied at the last meeting.

Lesson: Discuss each section of the chapter using the **HEADINGS IN CAPITAL BOLD FACED LETTERS** as a teaching outline. Ask students for questions or comments on what they have studied. Apply the lesson to the lives and ministries of your students.

Self-Test: Review the Self-Tests students have completed. (Note: If you do not want the students to have access to the answers to the Self-Tests, you may remove the answer pages from the back of each manual.)

For Further Study: You may do these projects on a group or individual basis.

Final Examination: If your group is enrolled in this course for credit, you received a final examination with this course. Reproduce a copy for each student and administer the exam upon conclusion of this course.

Module: Organizing
Course: Principles Of Environmental Analysis

COURSE INTRODUCTION

This course introduces principles of environmental analysis necessary for organizing and mobilizing spiritual resources for evangelizing.

This study stresses the importance of environmental analysis in planning strategies to reach the world with the Gospel. "Environmental analysis" is defined and reviewed in the Old and New Testament records. Instructions are given for doing a personal environmental analysis and analyzing people groups, geographic areas, nations, regions of the world, and Christian organizations.

This course is second in a series of three in the "Organizing Module" of training offered by Harvestime International Network. *"Biblical Management Principles"* precedes this course, and *"Management By Objectives"* follows it.

It is recommended that these three courses be studied in their suggested order for proper understanding of the leadership, planning, and organization necessary for effective ministry.

COURSE OBJECTIVES

Upon completion of this course you will be able to:

- Define the term "environmental analysis."
- Identify purposes for environmental analysis.
- Identify Old Testament examples of environmental analysis.
- Identify New Testament examples of environmental analysis.
- Do a personal environmental analysis.
- Analyze a geographic area.
- Analyze a nation.
- Analyze a region of the world.
- Analyze an existing Christian organization.
- Explain the "REAP" plan of Harvestime International Network.
- Use environmental analysis to plan strategies of evangelism.

CHAPTER ONE

THE ENVIRONMENT

OBJECTIVES:

Upon completion of this chapter you will be able to:

- Write the Key Verse from memory.
- Define the word "environment."
- Describe the first environment of man.
- Describe the present world environment.
- Explain what caused the change from the first to the present environment.
- Identify three ways the present environment affects man spiritually.
- Summarize God's plan for saving man from this sinful environment.
- Explain how the environment relates to the mission of the Church.

KEY VERSE:

And God saw that the wickedness of man was great in the earth, and that every imagination of the thoughts of his heart was only evil continually. (Genesis 6:5)

INTRODUCTION

Each born-again believer is called of God to become part of the great spiritual body known as the Church. The main commission given the Church is to take the Gospel to all nations:

Go ye therefore, and teach all nations, baptizing them in the name of the Father, and of the Son, and of the Holy Ghost:

Teaching them to observe all things whatsoever I have commanded you: and, lo, I am with you alway, even unto the end of the world. (Matthew 28:19-20)

...Thus it behoved Christ to suffer, and to rise from the dead the third day...

...That repentance and remission of sins should be preached in His name among all nations... (Luke 24:46-47)

Efforts to fulfill this challenge are affected by something known as "environment." This lesson introduces the subject of environment, defines the term, and explains how the environment relates to the mission of the Church.

ENVIRONMENT

The "environment" includes the physical, social, cultural, and spiritual factors which surround a person. It includes all parts of the society in which a person lives, works, and ministers.

The physical environment includes the visible material things which surround a man. It refers to the country, city, or village in which he lives. It refers to all that can be seen with the natural eye such as buildings, land, animals, trees, mountains, material possessions, etc.

The physical environment may be a dry dessert, cold mountain region, jungle, or island. It may be a very nice environment with mild weather where crops grow easily. It may be a very harsh environment, with bad weather which makes providing adequate food supplies difficult. The physical environment may be a crowded city populated with many people or a remote area with only a few people.

The social environment refers to the people who live in any physical environment. The social environment is the people among whom you live, work, and minister.

Individuals in every society are members of families. These families may be organized into castes, clans, or tribes. Within these groups are other divisions such as the poor, rich, educated, and uneducated. A nation is made up of many "people groups," about which you will study later in this course. The way people are organized in the environment is called the "social structure."

The cultural environment refers to the accepted patterns of life and behavior within a group of people. These patterns include such things as language, manners, customs, and material standards of living.

Culture also includes such things as the music, art, science, literature, educational systems and political structures. Culture develops as a result of religious and political influences and by repeated practice of customs over a long period of time.

Culture varies between nations and even between groups of people within a nation. For example, in Bible times the culture of Jews was different from that of the Gentiles. The Jews did not eat certain meats and they practiced circumcision. The Gentiles ate all kinds of meat and did not practice circumcision. The two groups varied in religion, customs, and social structure.

The spiritual environment refers to things affecting the spirit of man. God is a triune being: God the Father, God the Son Jesus Christ, and God the Holy Spirit. Man was created in the image of God as a triune being. Man has a body, soul, and spirit.

The spiritual environment is that which affects the soul and spirit of man. There are spiritual forces of evil under the control of Satan. There are spiritual forces of good under the control of God. These spiritual forces are constantly warring for control of the soul and spirit of man.

(Harvestime International Institute offers a course entitled *"Spiritual Strategies: A Manual Of Spiritual Warfare"* which covers the subject of the spiritual environment in detail.)

To understand the physical, social, cultural, and spiritual factors of man's environment, we must review the record provided by God in His written Word, the Holy Bible. We must go back in history to a period before time, as we know it, began here on earth.

THE BEGINNING OF ALL THINGS

The Bible teaches that God has always existed. He has no beginning and no end. His dwelling place is in Heaven. God created His own dwelling place in Heaven. He also created man's dwelling place, which is the earth. Everything that exists was created by God:

> **For by Him were all things created, that are in Heaven, and that are in earth, visible and invisible, whether they be thrones, or dominions, or principalities, or powers; all things were created by Him, and for Him.**
>
> **And He is before all things, and by Him all things consist. (Colossians 1:16-17)**

Everything God created was originally good:

> **And God saw everything that He had made, and behold, it was very good. (Genesis 1:31)**

THE ORIGIN OF EVIL

Among the living beings created by God were the angels. There was one special angel created by God named Lucifer. The Bible records that Lucifer was originally perfect, very beautiful, and held an important position in Heaven.

Lucifer became very proud of his beauty and position. He organized some of the other angels and rebelled against God. In pride, Lucifer wanted to be greater than God. He tried to overthrow God's Kingdom in Heaven.

Because of this sin, Lucifer and the angels who participated in his rebellion were cast out of Heaven to earth. Lucifer became known as Satan or the Devil. His angels became known as demons. You can read the history of these events in Isaiah 14:12-15 and Ezekiel 28:14-16. Satan is the spiritual force behind all sin, all acts of unrighteousness, and everything evil in the world.

THE FIRST ENVIRONMENT

When God first created our world, it was a perfect environment. God looked upon all the acts of His creation and said, "It is good" (Genesis 1). When God created man in His own image, man was also perfect. He was sinless and in spiritual fellowship with God (Genesis 1:26 and chapter 2).

When God created woman He told the first couple to reproduce and bring forth children. Adam and Eve were to populate the world with other sinless, godly people like themselves (Genesis 2:18-25; 1:28).

The first place Adam and Eve lived was a garden called Eden. This garden was a beautiful environment filled with the wonderful creations of God:

> **And the Lord God planted a garden eastward in Eden; and there He put the man whom He had formed.**
>
> **And out of the ground made the Lord God to grow every tree that is pleasant to the sight and good for food...**
>
> **And a river went out of Eden to water the garden... (Genesis 2:8-10)**

God created a perfect environment. But remember, Satan had been cast out of Heaven to earth. Satan caused a terrible change to come to this perfect environment in which man was placed by God.

SIN ENTERS THE ENVIRONMENT

Created in God's image, man had an intelligent mind with which he could make decisions. Man could choose to obey or disobey God. You will remember that Satan originated evil. Because evil existed, man could choose between good and evil. Man chose evil. The first disobedience of man is sometimes called "the fall of man," meaning that man fell from righteousness into sin. You can read the complete story of this tragic event in Genesis 2-3.

In the Garden of Eden, there were many trees from which Adam and Eve could eat. But there was one tree which the couple were to avoid:

> **And the Lord God commanded the man, saying, Of every tree of the garden thou mayest freely eat:**

But of the tree of the knowledge of good and evil, thou shalt not eat of it: for in the day that thou eatest thereof thou shalt surely die. (Genesis 2:16-17)

Man had the power of choice. He could obey or disobey God. Satan led the first man and woman to disobey God. Because of their disobedience, sin entered their environment.

THE ENVIRONMENT CHANGES

Sin brought terrible changes to man and his environment. God had warned Adam and Eve that the penalty of sin was death. This included both physical and spiritual death. Spiritual death was a loss of their relationship with God. Physical death was death of their actual physical bodies.

Because Adam and Eve sinned, sin and death came on all men:

Wherefore, as by one man sin entered into the world, and death by sin; and so death passed upon all men, for that all have sinned. (Romans 5:12)

Because of the fall of Adam and Eve, sin is passed on all men. That means everyone born inherits the basic nature of sin. Just as physical traits are inherited, the spiritual trait of the basic sin nature is inherited.

Each person sins individually when he is drawn by this sinful nature to rebel against God:

Every man is tempted when he is drawn away of his own lust and enticed.

Then when lust hath conceived, it bringeth forth sin: and sin, when it is finished, bringeth forth death. (James 1:14-15)

Because of the basic sin nature and individual acts of sin, all men face physical and spiritual death:

For all have sinned and come short of the glory of God. (Romans 3:23)

Man's physical environment was also cursed. God said the earth would bring forth thorns and thistles. Intense labor (sweat) marks man's relationship with his physical environment:

And unto Adam He said, Because thou hast hearkened unto the voice of thy wife, and hast eaten of the tree, of which I commanded thee, saying, Thou shalt not eat of it; cursed is the ground for thy sake; in sorrow shalt thou eat of it all the days of thy life.

Thorns also and thistles shall it bring forth to thee; and thou shalt eat of the herb of the field;

In the sweat of thy face shalt thou eat bread, till thou return unto the ground: for out of it wast thou taken; for dust thou art, and unto dust shalt thou return. (Genesis 3:17-19)

The very world itself became polluted by the sins of man:

...For thou hast polluted the land with thy whoredoms and with thy wickedness.

Therefore the showers have been witholden, and there hath been no latter rain. (Jeremiah 3:2-3)

It was not long until God looked upon the earth which He had created...

And God saw that the wickedness of man was great in the earth, and that every imagination of the thoughts of his heart was only evil continually. (Genesis 6:5)

Here is a summary of how man's environment changed because of sin:

1. **Man changed spiritually**: He was no longer in fellowship with God. "Spiritual death" means eternal separation from God because of sin. Everyone in future generations to come inherited the basic sin nature and all men have sinned.

2. **Man changed physically**: His body would eventually grow old, he would die, and he would return to the dust which God had used to create him.

3. **The earth itself was cursed**: "Thorns and thistles" represented the curse under which the environment was placed. Man would experience great difficulties between himself and this sinful environment in which he lived.

THE PRESENT ENVIRONMENT

The present environment of the world in which man lives is still evil. Until God destroys this present sinful world and creates the new Heaven and earth, all creation groans for release from the terrible environment created by sin:

For we know that the whole creation groaneth and travaileth in pain together until now. (Romans 8:22)

HOW MAN IS AFFECTED

The sinful environment of this present world affects man spiritually because three spiritual enemies are in operation: The world, the flesh, and the Devil:

1. **The world**: Satan works through the evil world system, with all its sin, materialism, and godless philosophies.

2. **The flesh**: All men are born sinners. Satan works through the basic sin nature to lead men and women to individual acts of sin. This basic nature (called the "flesh" in the Bible) results in terrible, sinful acts of unrighteousness.

3. **The Devil**: Satan is still at work in the environment to accomplish his purposes. He works through demons and men and women who submit their wills to him.

A WAY OF ESCAPE

God has provided a way of escape from this sinful environment. Man can be saved from the terrible effects of spiritual death (separation from God):

> **For God so loved the world that He gave His only begotten Son, that whosoever believeth in Him should not perish, but have everlasting life. (John 3:16)**

Through Jesus, man can also be saved from the eternal effects of physical death. All true believers who die physically will someday be resurrected from physical death to eternal life:

> **For the Lord Himself shall descend from Heaven with a shout, with the voice of the archangel, and with the trump of God: and the dead in Christ shall rise first:**
>
> **Then we which are alive and remain shall be caught up together with them in the clouds to meet the Lord in the air: and so shall we ever be with the Lord. (I Thessalonians 4:16-17)**

The sinful world in which we live will someday be destroyed, and God will create a new Heaven and a new earth:

> **But the day of the Lord will come as a thief in the night: in the which the heavens shall pass away with a great noise, and the elements shall melt with fervent heat, the earth also and the works that are therein shall be burned up...**

Nevertheless we, according to His promise, look for new heavens and a new earth, wherein dwelleth righteousness. (II Peter 3:10,13)

Even though the world is still under the curse of sin, believers can be freed from that curse and live abundant life through Jesus Christ. Jesus said:

...I am come that they might have life and have it more abundantly. (John 10:10)

Your basic sin nature can be changed by God's power to bear spiritual fruit of love, joy, peace, longsuffering, gentleness, goodness, faith, meekness, and temperance (Galatians 5:22-23). Even though you are still in the world, you do not have to be part of this evil sinful environment.

I pray not that thou shouldest take them out of the world, but that thou shouldest keep them from the evil.

They are not of the world, even as I am not of the world. (John 17:15-16)

You are in the world for the purpose of spreading the Gospel to this evil world:

As thou hast sent me into the world, even so have I also sent them into the world. (John 17:18)

We can rejoice because Jesus...

...gave Himself for our sins, that He might deliver us from this present evil world according to the will of God and our Father. (Galatians 1:4)

THE ENVIRONMENT AND THE MISSION OF THE CHURCH

The environment is important to the mission of the Church to reach all nations with the Gospel:

First: All ministry is done in the environment of a world which you have learned is basically evil. If you understand this you will not be surprised at the attacks of the enemy as you minister.

Second: There are many types of environments within this world which differ greatly one from another. Just as all men are affected by the evil environment of the world in general, they are also affected by the specific environment in which they live. When you understand their specific environment you can more effectively present the Gospel.

SELF-TEST

1. Write the Key Verse from memory.

____

____

2. Define the word "environment."

____

3. Describe the first environment of man.

____

4. Describe the present world environment.

____

5. What caused the change from the first environment to the present environment of the world?

____

6. Identify three ways the present environment affects man spiritually.

_______________ _______________

____

7. What is God's plan for saving man from this sinful environment?

____

8. How does the environment relate to the mission of the Church?

____

(Answers to tests are provided at the conclusion of the final chapter in this manual.)

FOR FURTHER STUDY

1. Study Genesis chapter 3 and observe how Satan used the environment in the first temptation. The Garden created by God was perfect. But there was one forbidden tree. Man was given a choice: To obey God and avoid this tree and continue to live in the perfect environment or to disobey God and let sin enter the environment. In Genesis chapter 3, Eve is in the environment near the forbidden tree. The Bible warns concerning evil to...

   Avoid it, pass not by it, turn from it, and pass away. (Proverbs 4:15)

   Eve heard: The voice of Satan from the environment of the forbidden tree.

   She saw: Something appealing in the environment. It was the forbidden fruit.

   She took: The sinful thing from the environment.

   She ate: The forbidden fruit. Before eating the evil was external. It did not affect her spiritually, but when she ate of it, she became sinful.

   She gave: Eve was part of the "social" environment of Adam because she was his wife. She influenced Adam to sin by giving him the forbidden fruit. People are often influenced to sin by those in their social environment.

   Adam ate: He also partook of the forbidden fruit. The cycle of sin had begun and would continue throughout history. All men became sinners and the environment of the world became sinful.

2. Describe the physical environment chosen by Lot in Genesis 13:8-11.

3. What kind of spiritual environment was in the city of Sodom which Lot chose? See Genesis 13:13 and chapters 18-19.

4. Compare the spiritual environment described in Genesis 6:5 to that described in Revelation 21:10-27.

CHAPTER TWO

ENVIRONMENTAL ANALYSIS

OBJECTIVES:

Upon completion of this chapter you will be able to:

- Write the Key Verse from memory.
- Define the term "environmental analysis."
- Identify six types of environmental analysis to be discussed in this course.
- List seven reasons why environmental analysis is important in taking the Gospel to all nations.
- Summarize six perils to avoid in environmental analysis.

KEY VERSE:

> **For whatsoever is born of God overcometh the world: and this is the victory that overcometh the world, even our faith. (1 John 5:4)**

INTRODUCTION

This chapter introduces "environmental analysis," a skill which leads to understanding the environments in which people live and how they affect efforts to reach them with the Gospel. "Environmental analysis" is defined and types of such analysis are identified. Perils to avoid are discussed and the importance of environmental analysis explained.

DIFFERING ENVIRONMENTS

In the last chapter you learned that the environment includes the physical, social, cultural, and spiritual factors which surround a person. You also learned that the world is sinful and inhabited by sinful men and women in need of a Savior.

All ministry is done in the general environment of this world. But there are many varied environments within this sinful world and each one differs greatly from another. For example, the environment of a small tribe in the Sahara Dessert differs greatly from that of a resident of a modern city like New York.

Just as all men are affected by the evil environment of the world in general, they are also affected by the specific environment in which they live. When you understand their environment, you can more effectively present the Gospel message.

ENVIRONMENTAL ANALYSIS

"Environmental analysis" will help you understand a specific environment. To analyze something means to study it in detail, examining each part that makes up the whole. "Environmental analysis" means to study each part of the environment including physical, social, cultural, and spiritual factors. You analyze a specific environment in detail.

TYPES OF ENVIRONMENTAL ANALYSIS

There are six types of environmental analysis discussed in this course which are important to the mission of spreading the Gospel to the world. Each will be covered in detail in later chapters:

PERSONAL ANALYSIS: This is analysis of your own environment and yourself as a believer.

ANALYZING A PEOPLE GROUP: The world is made up of many different people groups. These groups speak different languages, have different customs, and are organized in various ways. Each geographic area, nation, and region of the world is made up of many different people groups.

ANALYZING AN AREA: An area is a specific geographic area within a nation. It may be a county, city, village, province, territory, or state.

ANALYZING A NATION: A nation is all the land and people joined together by political boundaries living under similar government.

ANALYZING A REGION: A region of the world is made up of several nations in a specific part of the world. For example, the Central American region includes the nations of Guatemala, Costa Rica, Nicaragua, Belize, Panama, Honduras, and El Salvador.

ANALYZING AN EXISTING CHRISTIAN ORGANIZATION: These are churches, denominations, and organizations that have been established for the purpose of Christian ministry.

THE IMPORTANCE OF ENVIRONMENTAL ANALYSIS

Understanding the environment is important to the spread of the Gospel because it...

HELPS BELIEVERS DISCOVER THEIR PLACE IN MINISTRY:

In the lesson on "Personal Environmental Analysis," you will analyze your spiritual strengths, weaknesses, calling, and gifts. This analysis is important because it helps you discover your specific place in ministry. It identifies spiritual weaknesses with which you must deal in order to be more effective and spiritual strengths which you can use for God.

AFFECTS EVERY PERSON:

In spreading the Gospel, it is important to understand that no one escapes the effects of the evil environment of this present world. All men of every culture, tribe, and nation are sinners in need of a Savior. There is evil present in every culture and the basic sin nature is active in every person who is an unbeliever. Satan will be in operation in every environment of this world in which you minister. You will continually war against the world, the flesh, and the Devil.

RELATES THE GOSPEL TO THE ENVIRONMENT:

The ability of people to understand the Gospel is affected by their environment. For example, you must share God's Word in a language that is understood by the listener.

You must also share the Gospel in a manner that is culturally and socially accepted. In some people groups it is necessary to gain approval and acceptance of tribal leaders before the people will even listen to your message. The way you present your message is affected by the social and educational levels of your audience. You must adjust the message accordingly.

You may have to adjust your life style to be accepted by a certain culture. For example, the Apostle Paul ate meat, but some cultures in which he ministered did not. He adjusted his life style in order to reach these people with the Gospel:

> **Wherefore if meat make my brother to offend, I will eat no flesh while the world standeth, lest I make my brother to offend. (I Corinthians 8:13)**

You must NEVER adjust your life style to conflict with God's written Word, but where simple personal changes can be made in order to advance the Gospel it should be done.

In many societies you can effectively use the culture to communicate the Gospel. But to do so, you much understand the cultural environment of the people. Culture can be used as a bridge into the Kingdom of God or a barrier that keeps people out.

ENABLES USE OF THE ENVIRONMENT TO TEACH THE GOSPEL:

Jesus used the environment to teach the Gospel. He told stories which related to the culture of the people by using common objects like wheat, fish, flowers, candles, and buildings to teach spiritual lessons.

Jesus did not remove His students to a false environment. He did not shut them away in a convent or monastery. Jesus taught men and women right where they lived, worked, and ministered. He related the Gospel to their personal situations. The environment was the training ground in which they applied what they learned.

(This is the vision of Harvestime International Network. This is why this course comes to you right in your own environment.)

IDENTIFIES FERTILE SPIRITUAL GROUND:

In the natural world, farmers analyze soil before planting their crops. They do this for three basic reasons:

First, they want to select the very best soil to concentrate their efforts. They know that the most fertile ground will bring the greatest harvest.

Second, they identify poor soil to determine how it can be prepared for farming. In some cultures farmers use commercially produced fertilizers to improve the soil. Others cut trees and foliage, burn it on the land, and use the ash created as a natural fertilizer.

Third, they analyze the soil to determine the best methods of farming. For example, do they need to use methods of irrigation or is rainfall sufficient? What will they plant? How will they harvest the crop?

Environmental analysis in the spiritual world is similar to what a farmer does in the natural world. We analyze people, areas, nations, and regions for similar reasons:

First, we want to determine the most fertile areas spiritually. Some areas are ready for spiritual harvest. The Bible says laborers are few in God's harvest fields. Since the spiritual harvesting force is limited, we want to concentrate our main efforts in areas ready to be reaped spiritually.

Second, we want to determine methods to make spiritually unreceptive areas ready for harvest. Some peoples, areas, nations, and regions are not as receptive to the Gospel. We must discover why and what methods will help change this spiritual soil.

Third, we determine the best methods to use...and that leads us to another major reason for analysis which is that it...

IDENTIFIES FACTORS WHICH AFFECT YOUR MISSION:

The mission of the Church is to spread the Gospel. The environment affects the strategy by which this mission will be accomplished. "Strategy" is a method or plan for achieving a goal. If you are to achieve the goal of evangelizing the world with the Gospel, then you must have some "strategies" or methods to achieve this goal. (You will learn more about strategies and planning in the last chapter of this course.)

The same plan will not work for every person the same way in every place. This is because no two people and no two environments are exactly alike any more than no two crops or fields are exactly alike in the natural world.

When you analyze people groups, geographic areas, nations, and regions, you obtain information that helps you plan strategies to effectively reach people in these different environments with the Gospel. The environment determines acceptable methods for sharing the Gospel because of differing political systems.

For example, in some countries you can no longer conduct open air evangelistic meetings. In some nations you cannot obtain land to build a church. In other nations you can no longer obtain a missionary visa to enter the country to preach the Gospel.

Such regulations do not mean you give up in defeat. You simply adjust your methods of sharing the Gospel in that environment. For example, instead of going in as a missionary, you might obtain a visa to work as a secular laborer while at the same time sharing the Gospel. Instead of open air meetings, you might use methods of personal evangelism. Instead of meeting in church buildings, you might raise up congregations that meet in homes of believers.

Analyzing the environment helps you plan details such as locating churches and services where people can get to them and timing them at an hour when people are able to attend. Analyzing the environment helps you identify strategies for sharing the Gospel that will be effective and accepted physically, culturally, socially, and spiritually.

RELATES TO GOD'S DEALINGS WITH MAN:

God sent a great flood on the earth to judge man because of the terrible environment resulting from his sin:

> **And God saw that the wickedness of man was great in the earth, and that every imagination of the thoughts of his heart was only evil continually.**
>
> **And it repented the Lord that he had made man on the earth, and it grieved him at his heart.**
>
> **And the Lord said, I will destroy man whom I have created from the face of the earth... (Genesis 6:5-7)**

Read the story of the Sodom and Gomorrah in Genesis 18-19. God sent judgment on these cities because of the sinful environment they had created.

Read the story of Ninevah in the book of Jonah. God planned to send judgment upon this city because of sin. When the entire city repented, God dealt differently with them because the environment of the city and their personal lives had changed.

There are many other Biblical examples of how the environment and God's dealings with man are related. If you are to understand the dealings of God with man, you must understand the environment.

PERILS TO AVOID

Environmental analysis is a Biblical concept. You will study in detail its Old and New Testament basis in the next two lessons. But before proceeding further with this study, there are some perils of which you should be aware.

To identify these perils, we will use the story of an environmental analysis done by twelve men in the Bible. First read the story in Numbers 13-14.

There are some specific perils of environmental analysis illustrated in this story. Here are some perils to avoid:

DO NOT ANALYZE WITH NATURAL REASONING ONLY:

Much of the information you obtain in environmental analysis is gained by studying and observing things in the natural world. But you must look beyond the natural world into the spiritual world.

The spies who went into Canaan saw giants in the natural world. Where they made their mistake was in analyzing the environment with natural reasoning only. Natural reasoning saw the giants as too great.

Do not make spiritual decisions on the basis of natural facts alone. Natural circumstances may be studied to guide your methods, but things must be analyzed spiritually also.

DO NOT BE OVERWHELMED BY THE TASK:

Ten of the spies who studied Canaan were overwhelmed by the greatness of the task. They saw the giants as too big to conquer. They were looking at the enemy's power and their own inabilities. They were focusing on obstacles instead of the opportunities.

An analysis of a specific environment may reveal many obstacles. But do not focus on the greatness of the task. Focus on the greatness of God. Do not look at your inabilities, but focus on God's abilities. Do not be overwhelmed by the power of the enemy because greater is He that is in you than he that is in the world (I John 4:4 and 5:4).

DO NOT BE INFLUENCED BY NEGATIVE REPORTS:

The negative report given by ten of the spies in Numbers 13 influenced the decision of God's people. When you prepare to do a work for God there will always be those who tell you it cannot be done. They will fill you with negative thoughts and reports.

Do not let them influence you. If you do, you may find yourself wandering in a spiritual wilderness as did Israel. You may miss the wonderful opportunity of entering and claiming new territory for God.

DO NOT BE FEARFUL AND UNBELIEVING:

The Israelites were kept out of their promised land of Canaan because of fear and unbelief resulting from their analysis:

> **But with whom was He grieved forty years? Was it not with them that had sinned, whose carcasses fell in the wilderness?**
>
> **And to whom sware He that they should not enter into His rest, but to them that believed not?**
>
> **So we see that they could not enter in because of unbelief. (Hebrews 3:17-19)**

Do not be kept out of any area because of fear or unbelief resulting from what you learn about it.

DO NOT GATHER FACTS JUST FOR FACTS SAKE:

You do an analysis for a purpose, not just for the sake of obtaining facts about an environment.

The spies who went into Canaan gathered many facts about the environment, its people, and its products. But they never acted upon what they learned. They never moved in to possess the land for God.

You can spend all your time gathering information and planning, but unless you get out into the harvest field and apply your strategies, people will not be reached with the Gospel.

DO NOT FOCUS ON "IF," BUT RATHER ON "HOW":

In Numbers 13 the spies who were sent into Canaan were sent to determine HOW to best take the region, not to decide IF they should take it. Forty years later when Israel came again to the border of Canaan, spies were again sent in to analyze Canaan. But these men understood their mission better than the first spies. They were to determine HOW, not IF, they were to proceed to take the land for God (Joshua 2).

As believers, we are commissioned to reach all nations with the Gospel. When we conduct an analysis it is for the purpose of determining HOW best to reach any given peoples with the Gospel.

SELF-TEST

1. Write the Key Verse from memory.

____

2. Define the term "environmental analysis."

____

3. Identify six types of environmental analysis you will study in this course.

_________________________ _________________________

_________________________ _________________________

_________________________ _________________________

4. List seven reasons why environmental analysis is important to the commission of the Church to spread the Gospel to all nations.

_________________________ _________________________

_________________________ _________________________

_________________________ _________________________

____

5. Identify six perils to avoid in environmental analysis.

_________________________ _________________________

_________________________ _________________________

_________________________ _________________________

(Answers to tests are provided at the conclusion of the final chapter in this manual.)

FOR FURTHER STUDY

1. Paul used the environment as a bridge instead of a barrier to the Gospel. Study the following references to see how the environment affected his strategy (methods) for sharing the Gospel:

   -He ministered to those of "low" estate by approaching them at their level: Romans 12:16

   -He used language appropriate to the culture: Acts 22:1-2

   -He adjusted His behavior so it would not offend cultural or religious beliefs: I Corinthians 8:10-13, Romans 14:15

   -He did not impose His Jewish culture on those accepting the Gospel: Acts 15:1-32

2. An example of what happens when you analyze only on the basis of natural reasoning is recorded in Acts 27. The crew of the ship analyzed things with natural reasoning only. Their analysis made it appear that they should proceed. But Paul had received spiritual insight. God had revealed it was not the proper time to proceed with the trip.

3. An example of being overwhelmed by the circumstances identified in an analysis is given in II Kings 6:14-17. Gehazi focused on the great task instead of the greatness of God.

CHAPTER THREE

THE BIBLICAL BASIS: OLD TESTAMENT

OBJECTIVES:

Upon completion of this chapter you will be able to:

- Identify Old Testament examples of environmental analysis.

INTRODUCTION

The Bible is filled with examples of environmental analysis. It is a record of man's environment on earth and his response to the influences of that environment. For example, the Bible analyzes the past, present, and future environments of man. It analyzes the spiritual condition of men and women apart from God and contrasts it with that of true believers. It also analyzes our spiritual enemies.

In this and the following lesson you will study Biblical examples of environmental analysis. It is impossible to analyze every reference relating to the subject because there are so many. For this reason only key passages have been selected as examples.

(Due to the study format used in this lesson, there is no "Key Verse," "Self-Test," or "For Further Study" section in this chapter.)

GENESIS

Chapters l-2: God creates a perfect environment.

Chapter 3: Sin enters the environment. How was it affected? How did God analyze the spiritual condition of Adam and Eve after their sin?

6:5-6: Note how the environment had changed from that described in Genesis l-2.

Chapters 6-7: God analyzes and deals with the wicked environment on earth.

Chapter 11: Read the story of the tower of Babel. Compare the analysis done by the people to that of God.

13:6-13: Decisions determine destiny. Lot made a wrong choice based on an analysis of the environment done by natural reasoning instead of spiritual wisdom.

Isaac: He was not limited by the challenges of his environment. (See 24:34-35; 25:32; 26:12-14.)

Chapter 22: Abraham analyzed an impossible situation in the natural world and decided to take things in his own hands rather than wait for God's plan.

Chapters 37-48: By divine revelation Joseph analyzed the present and future environments and saved an entire generation from famine.

EXODUS

Chapter 2: Moses's mother analyzed a dangerous environment and saved her son's life.

Chapters 2-3: Moses analyzed his environment, responded wrongly, then fled to the desert.

6:9: Note how the environment influenced the response of the Israelites to God's message.

Chapters 7-13: If Pharaoh had correctly analyzed the environment, he would have recognized God's dealings with him and properly responded.

Chapter 14: Analyzing the environment with natural reasoning, Israel was stranded. God intervened and they were saved. We cannot analyze situations on the basis of natural reasoning alone.

15:22-27: Israel at the waters of Marah. What was their response to the environment? How was the environment used by God?

Chapter 32: Aaron blames the people (his environment) for his sin.

18:1-27: Jethro's analysis of the environment resulted in new organization.

LEVITICUS

The book of Leviticus is a record of God's laws for His people. Many of the laws relate to the natural environment, i.e., dietary, health, and sanitation situations. Others relate to the political environment, i.e., how to handle disputes and problems. Others relate to the spiritual environment, i.e., God's plan for dealing with their sins.

NUMBERS

Chapters 13-14: A contrast of spiritual and natural analysis. Ten spies viewed the land through natural reasoning. Two analyzed the situation through spiritual insight.

Chapter 16: Korah analyzed a situation and made a bad decision. What was wrong with his analysis?

Chapters 22-25: Read the story of Balak and Balaam. Balak's analysis resulted in his appeal to Balaam.

Note: The books of Exodus, Numbers and Deuteronomy relate events during the wandering of Israel in the wilderness after their refusal to enter the Promised Land because of fear and unbelief. Much can be learned by the way God's people analyzed and dealt with their environment during this 40 year journey.

DEUTERONOMY

Chapters 1-4: These chapters present a historical review of Israel's wilderness journey. Note how the environment and their response to it affected their relationship with God.

Chapter 6: How can a good and prosperous environment have a negative effect on your relationship with God?

JOSHUA

Chapter 2: Environmental analysis done by two spies. Note how their analysis differed from that of the first spies in Numbers 13-14. Faith never minimizes the difficulties after analyzing them. Faith just sees them as being small in comparison to God's power.

Chapter 7: Joshua was not alert to the spiritual environment and suffered defeat.

Chapter 9: The results of analyzing in the natural senses without the counsel of God. Note verse 14.

Note: Environmental analysis is important throughout the book of Joshua in preparation for the various battles to possess the land. The battle plan was usually based on the result of analyzing the environment. This is true of many of the Old Testament battles. Can this be applied spiritually?

JUDGES

The book of Judges is a historical record of kings who ruled God's people. Some kings were good and others were evil. The book also records the stories of godly judges who were raised up by God to deliver His people from bondage. The influence of godly and ungodly leadership on people and their environment is stressed. How God's people responded to these influences in their environment determined how God responded to them spiritually.

RUTH

Read through this short book and note how Naomi analyzed her situation in discouragement while Ruth responded in faith (chapter 1). Note also how Naomi analyzed and responded to the situation with Boaz and his analysis of Ruth's situation. Boaz worked within the culture of his environment in redeeming Ruth (chapter 4).

I SAMUEL

2:11-3:21: Eli did not properly analyze the spiritual condition of his own household.

15:1-35: Saul analyzed on the basis of natural wisdom rather than God's commands. He lost the kingdom because of it.

II SAMUEL

Chapter 12: A powerful spiritual analysis by Nathan the prophet. It resulted in David repenting from sin.

Chapters 13-19: The story of Absalom's rebellion. How was he influenced by his environment? Did David refuse to see the situation as it really was?

I KINGS

9:1-10:29: What was Solomon's natural environment like?

11:1-43: How did the environment affect Solomon spiritually?

II KINGS

This book is a record of the various kings who ruled God's people. Note how the environment was affected by godly and evil leaders. How were the people affected?

I AND II CHRONICLES

The religious history of God's people is recorded in these books. As you read these books, observe how the environment was affected by spiritual and evil leaders.

EZRA

Ezra analyzed the environment of Jerusalem and realized the need for rebuilding the Temple. During the rebuilding, he encountered much opposition from those in his environment. With spiritual insight, he analyzed each situation and responded properly. Note especially his analysis and response to the sins of God's people in chapters 9-10

NEHEMIAH

Environmental analysis is stressed in this book which concerns the rebuilding of the wall around Jerusalem by a man named Nehemiah. Especially note in 2:11-20 his analysis of the situation in Jerusalem. Throughout the book, Nehemiah continues to analyze the environment and respond in faith towards the negative factors which could have hindered his work for the Lord.

ESTHER

Chapter 2: Mordecai's analysis of a situation resulted in Esther being in the right place to be used of God.

Chapters 3-9: Esther analyzes a situation and formulates a plan to save an entire nation.

JOB

This book records the story of a righteous man named Job who was attacked by Satan physically, mentally, materially, socially, and spiritually. Job analyzes his own situation, his friends conduct improper analysis based on natural reasoning, and the book climaxes with God's analysis.

PSALMS

1:1-3: The environment of the godly man is analyzed.

3:6: Do not fear forces in the environment.

11:4-5; 14:2-3; 17:3: God analyzes all creation.

19:12,14; Chapters 26, 51, 69; 40:2-3; 119:59: David spent much time analyzing himself. These are just a few examples of many found in the Psalms.

Chapters 78, 105-107: Analysis of the nation of Israel.

115:4-8; 135:15-18: An analysis of false idols.

PROVERBS

Many passages in this book analyze and warn against the environment of the wicked. (See 1:10-19; 2:10-22; 4:14-19; 7:6-27; 9:13-18.)

Chapter 5; 14:7: We should flee from some environments. (Note especially 5:8.)

6:6-11: Study the environment of the ant and see what you can learn.

6:27-28; 22:24-25: The environment, both natural and spiritual affects us.

20:4: Do not let your environmental analysis prevent you from doing your work.

21:2: God analyzes our hearts.

21:29; 4:26: We are told to "consider your ways" and "ponder the path of your feet." This is personal analysis.

24:30-34: The field of the slothful is analyzed.

27:23: "Know the state of thy flocks" means we are to be aware of those for whom we are responsible.

ECCLESIASTES

This book is an analysis of life "under the sun" on earth apart from God. The writer (Solomon) analyzes everything in the environment of the present world and finds it all vanity. The conclusion of his analysis is given in Ecclesiastes 12:13-14.

SONG OF SOLOMON

This book shows the relationship between Christ and the Church as illustrated by the marriage relationship. Note the personal analysis of the bride and bridegroom.

ISAIAH-DANIEL

The books of Isaiah through Daniel are called "major prophets." They contain lengthy prophecies given to various nations by prophets under the inspiration of God. Review these books and note the descriptions of the sinful environments in which the people of these nations lived. Note the prophecies of how God planned to deal with them because of these sinful environments. Are there things in the environments of these nations that you see evident in the world today?

Note also the story of Daniel and his three friends in chapters 1-3 and 6. They did not allow their sinful environment to affect their spiritual relationship with God.

HOSEA-MALACHI

The books of Hosea through Malachi are called "minor prophets" because of their shorter length, not because their message is any less important.

These prophets, under the inspiration of God, analyze the environment of various nations. Review these books, noting the descriptions of the sinful environment of these nations and how God planned to deal with them because of it. Are there things in the environments of these nations that you see evident in our world today? Note Amos 4:6-13 which explains how God used the environment to deal with His own people. If they had properly analyzed their environment and repented, God would have responded in forgiveness and healing.

CONTINUE YOUR STUDY...

This brief review of the Old Testament illustrates the importance of understanding and properly responding to the environment. God's people...

...Were defeated when they analyzed the environment with attitudes of fear and unbelief or when they failed to analyze at all.

...Won great victories when they analyzed the environment with spiritual insight.

...Allowed their environments to influence them spiritually, both positively and negatively.

...Affected the environments in which they lived, worked, and ministered.

...Worked within the culture of their environments to accomplish God's plan.

There are many other examples of environmental analysis in the Old Testament. Continue this study and expand this lesson with your own personal notes on the subject.

CHAPTER FOUR

THE BIBLICAL BASIS: NEW TESTAMENT

OBJECTIVES:

Upon completion of this chapter you will be able to:

- Identify New Testament examples of environmental analysis.

INTRODUCTION

In the last chapter you studied the Biblical basis of environmental analysis in the Old Testament. In this lesson, you will review its New Testament basis.

(As in the last chapter, because of the study format used in this lesson, there is no "Key Verse," "Self-Test," or "For Further Study" section.)

Here are some New Testament examples of environmental analysis:

THE GOSPELS

Matthew, Mark, Luke, And John

The Gospels are similar accounts of the environment of the time of the ministry of Jesus and of His use of environmental analysis. For this reason, these books are considered together. Focus on the ministry of Jesus in the Gospels. Note specifically how Jesus...

...Analyzed the environment to identify and meet the needs of people.

...Analyzed the questions of His critics so He would not be tricked.

...Analyzed the thoughts and actions of people. (For examples, the young man who did not respond to the call of Jesus because of riches; the motives and plans of Judas; the true level of Peter's commitment, etc.)

...Analyzed the present conditions of cities in which He ministered. (For examples, see Matthew 23:37-39; Luke 13:34-35; and Matthew 11:21-24.)

The Gospels provide an analysis of the religious environment at the time of Christ. What is revealed about the religious leaders based on information given about the Scribes, Pharisees, and Sadducees?

Here are some other miscellaneous references concerning environmental analysis:

Matthew 7:15-20: How to analyze true and false ministries.

Matthew 13: Analyzing for receptive ground.

Matthew 10:5-15; Mark 6:10-11: Analyze the environment when you minister and respond accordingly.

Mark 13:5-37: An analysis of future conditions on the earth.

Luke 14:7-14, 28-32: Examples which stress the importance of environmental analysis.

ACTS

2:41-47; 5:42: These verses analyze the spiritual environment of the first Church at Jerusalem.

6:1-6: An analysis of conditions which resulted in the selection of the first deacons.

6:7: An analysis of Jerusalem.

8:17-24: Peter's analysis of a man named Simon.

10:34-11:18: Peter's analysis of God at work among the Gentiles.

11:27-30: An analysis of conditions in Judaea.

Chapters 13-28: In these chapters, Paul visited many cities and ministered to many different types of groups. Note how Paul analyzed and adapted the Gospel to the various environments.

Chapter 27: Two opposite analyses and the results.

ROMANS

1:7-8: Paul analyzes Roman believers. What three things does he say about them?

1:21-32: A description of the environment of evil men who have fallen short of God's glory.

3:9-19, 23; 5:12: Analysis of a world without God.

10:1-3,21: Analysis of the nation of Israel.

11:2-4: An analysis by the prophet Elias is recalled.

16:17-18: An analysis of those who cause division in the Church.

I CORINTHIANS

Chapter 8: Paul analyzes false gods (idols).

9:19-22: Paul's personal analysis.

10:27-33: Analyze your environment carefully. Do not offend others by your actions.

16:8-9: An analysis of the city of Ephesus.

II CORINTHIANS

1:8-9: An analysis of Asia.

2:12-13; 7:5; 8:1-5; 11:9: An analysis of Macedonia.

4:7; 5:17: An analysis of believers in Christ.

9:1-4: Analysis of Achaia.

Paul analyzes the Church at Corinth in the books of I and II Corinthians. What do you learn about the Church from his analysis?

GALATIANS

1:6-7; 3:1-5; 4:9-20; 5:1,7-8; 6:15: What is revealed about the Church at Galatia based on information given in these verses?

1:4: An analysis of the present world.

1:11-2:21: A self-analysis by Paul.

Chapter 3: An analysis of Abraham.

5:16-21; 6:7-8: Analysis of the life controlled by flesh.

5:22-26: Analysis of a life controlled by the Spirit of God.

6:3: Be careful how you analyze yourself!

EPHESIANS

1:3: Paul's analysis of the spiritual blessings in Christ which belong to the believer. He identifies many of these in the book of Ephesians. Make a list of these, based on Paul's analysis in these chapters.

2:1-3, 11-12: Analysis of what we were like in "times past" before we knew the Lord.

2:4-10, 13-22: What are we like now that we know Jesus?

3:7-8; 4:1; 6:19-20: A personal analysis of Paul.

4:3-6: Analysis of the one true Body of Christ.

4:11-16: An analysis of the purposes of spiritual gifts.

4:17-5:9: Study Paul's analysis of our conduct in "times past." Contrast it with instructions for our conduct as believers.

6:11-18: What does this analysis reveal about our true spiritual enemies?

PHILIPPIANS

1:12-14,19: Paul's analysis of his circumstances.

1:15-18: An analysis of various motives for ministry.

1:20-25: Analyzing benefits of life and death.

2:5-8: An analysis of the mind of Jesus.

2:25-30: Paul's analysis of his friend, Epaphroditus.

3:3-14: Paul's personal analysis.

3:18-21: An analysis of the enemies of the cross.

4:11-13: How Paul analyzed and dealt with his circumstances.

4:14-18: What can you learn about the Church at Philippi based on the information Paul provides in these verses?

COLOSSIANS

In the Colossian Church was a group of professing believers who brought in false teaching. What does Paul's analysis tell you about the situation in the Church at Colosse? How does his analysis help him deal with the situation there? What commands does Paul give to correct the situation and guard against false doctrine?

I AND II THESSALONIANS

What can you learn about the Thessalonian Church based on the information given in these books? What were some good things about these congregations which Paul commends? In what areas did they need improvement?

I TIMOTHY

I and II Timothy were both written to a young minister named Timothy. Read both books and observe what Paul analyzed as necessary qualities to be a successful minister of the Gospel.

1:4-7: An analysis of what caused certain men to be swayed from the true Gospel.

1:8-10: An analysis of the purpose of the law.

1:11-16; 2:7: A self-analysis by Paul.

1:18-20: What can you learn about Hymenaeus and Alexander from the analysis given by Paul in these verses?

3:1-13: An analysis of requirements for leaders in the Church.

4:1-5: An analysis of conditions of the last times.

4:6-6:21: Study Paul's analysis of things which will make Timothy a good minister God.

II TIMOTHY

1:5-6, 13-14; 3:15: Based on the information given in these verses, what can you learn about Timothy's environment.

1:11-12; 2:9-10; 3:10-11; 4:6-8,18: A self-analysis by Paul.

1:15-18: An analysis of conditions in Asia and of a man named Onesiphorus.

3:1-9,13; 4:3-5: An analysis of the last times.

3:16-17: An analysis of the purposes of God's Word.

4:9-17: An analysis of different men to whom Paul related.

TITUS

The book of Titus was also written by Paul to a young minister. What does Paul analyze as necessary to be a successful minister of the Gospel? In addition, study these passages also:

1:9-16; 3:10-11: An analysis of ungodly men.

3:3-7: An analysis of what we "were" compared to what we "are now" in Christ.

PHILEMON

Read this short book of one chapter. Can you analyze the situation with which Paul was dealing on the basis of the information provided in these verses?

HEBREWS

Chapters 3-4: An analysis of why Israel could not enter into the good things of God.

5:12-14: An analysis of believers who are unskilled in God's Word.

Chapter 7: An analysis of the struggles with life in the flesh.

Chapter 8: A contrasting analysis of life in the Spirit.

Chapter 11: An analysis of faith and its importance in the lives of great men and women of God.

Chapter 12: An analysis of the purposes of chastisement.

The book of Hebrews presents the old and new covenants of God with man, analyzing these different plans of His relationship with man.

JAMES

1:6-8: An analysis of a double minded man.

1:12-14: An analysis of environments of temptation.

1:17: An analysis of the source of good gifts.

1:27: An analysis of the basics of pure religion.

2:1-9: An improper use of analysis.

2:20-24: A personal analysis of Abraham.

2:25-26: A personal analysis of Rahab.

3:1-13: A frightening analysis of our own tongue which affects our environment.

3:14-16: An analysis of environments where confusion is evident.

4:1-3: An analysis of conditions which result in wars and fighting.

5:1-6: An analysis of rich men.

5:16-18: A personal analysis of Elias.

I PETER

Peter deals with the environment of suffering in these chapters. What does he reveal in his analysis of its purpose, results, and the proper attitude of believers in such environments? Read the book again and take note of instructions given to believers to create a proper environment in the Church through godly living.

II PETER

Chapters 2-3: An analysis of certain men who were deceivers. Compare Peter's analysis to that of Jude.

I JOHN

John lists many ways you can analyze yourself to determine if you really know God. Read through the book of I John and list these.

1:6-10: The importance of self-analysis.

2:15-17; 5:19: An analysis of the world.

2:18-19: How to analyze and recognize the spirit of those who are anti-Christ.

4:1-6: Analyze spirits on the basis of these instructions.

5:4: How to overcome any environment.

II JOHN

Verse 4: An analysis of the believers to whom he was writing.

Verses 7-11: An analysis of deceivers.

III JOHN

Verses 1,3-8: An analysis of a man named Gaius.

Verses 9-12: An analysis of the differing examples of Demetrius and Diotrephes.

JUDE

Verse 4: An analysis of "certain men" begins, on which the entire chapter focuses. What do you learn about these men from study of this chapter?

Verse 5: An analysis of Egypt.

Verse 6: An analysis of wicked angels.

Verse 7: An analysis of Sodom and Gomorrah.

REVELATION

The book of Revelation is an analysis of conditions on earth during the final days of time as we now know it. Read through the book and note these conditions.

Chapters 2-3: An analysis of the environments of the seven churches of Asia. Can you see any of these qualities in the modern Church?

Chapters 21-22: An analysis of the environment of the new Heaven and earth which God will create.

CHAPTER FIVE

PERSONAL ENVIRONMENTAL ANALYSIS

OBJECTIVES:

Upon completion of this chapter you will be able to:

- Write the Key Verse from memory.
- Define "personal environmental analysis."
- Explain what a personal environmental analysis includes.
- Identify purposes of a personal environmental analysis.
- Do a personal environmental analysis.
- Use the results of your analysis to fulfill your ministry.

KEY VERSE:

> **Examine yourselves: whether ye be in the faith; prove your own selves...**
> **(II Corinthians 13:5)**

INTRODUCTION

You have received an introduction to environmental analysis and examined its Biblical basis. With this chapter you begin a series of lessons which explain how to analyze various environments. Be sure to complete each study before you proceed to the next as the lessons build one on another. In this lesson you will learn how to do a personal environmental analysis.

PERSONAL ENVIRONMENTAL ANALYSIS

A personal environmental analysis is a study of your own environment. It includes an analysis of every area of your life including your relationships and your physical, educational, financial, and spiritual environments.

THE PURPOSE OF PERSONAL ANALYSIS

Here are the purposes of personal analysis:

TO DETERMINE YOUR SPIRITUAL CONDITION:

The Bible says you are to examine yourself to determine your own spiritual condition:

Examine yourselves: whether ye be in the faith; prove your own selves... (II Corinthians 13:5)

But let a man examine himself... (I Corinthians 11:28)

You are to examine yourself in relation to the "faith." This means to analyze your spiritual condition on the basis of God's Word. "Prove" is another word for examine. The Bible says:

Prove all things; hold fast that which is good. (I Thessalonians 5:21)

Self-examination includes analyzing the spiritual areas of your life, but it also includes analysis of practical areas. These include such things as your physical condition, finances, education, etc. You must "prove" or examine "all things" in your life.

Man has a body, soul, and spirit. What happens to the body affects your spiritual condition. The condition of the soul and spirit can affect the spiritual body. This is why you examine "all things."

It is also important to examine every area of life because practical concerns of life can interfere with God's plan for you. Things of the world can hinder the work of God's Word in your life:

And the cares of the this world, and the deceitfulness of riches and the lusts of other things entering in, choke the Word, and it becometh unfruitful. (Mark 4:19)

Practical concerns of life can hinder spiritual fruitfulness:

And that which fell among thorns are they, which, when they have heard, go forth, and are choked with cares and riches and pleasures of this life and bring no fruit to perfection. (Luke 8:14)

Involvement in details of living can hinder spiritual warfare:

No man that warreth entangleth himself with the affairs of this life; that he may please Him who hath chosen him to be a soldier. (II Timothy 2:4)

Practical concerns of life can so overcome you that you will not be properly prepared for the Lord's return:

And take heed to yourselves lest at anytime your hearts be overcharged with surfeiting, and drunkenness, and CARES OF THIS LIFE, and so that day come upon you unawares. (Luke 21:34)

"Take heed to yourselves" means to examine yourself to be sure that in every area of life you are ready for the Lord's return.

TO IDENTIFY YOUR SPIRITUAL WEAKNESSES:

Self-improvement is not the purpose for identifying your weaknesses. Read Romans chapter 7 to see how Paul tried this and failed. You can do nothing by yourself to change your weaknesses.

The purpose of identifying your weaknesses is to permit you to cooperate with God's process of change in your life. Change starts with the new-birth experience:

> **Therefore if any man be in Christ, he is a new creature; old things are passed away; behold, all things are become new. (II Corinthians 5:17)**

Change continues by confessing your sins and weaknesses and accepting God's forgiveness:

> **If we say that we have no sin, we deceive ourselves, and the truth is not in us.**
>
> **If we confess our sins, He is faithful and just to forgive us our sins and to cleanse us from all unrighteousness. (I John 1:8-9)**

You change as you cooperate with God in the process of changing you through the circumstances of life:

> **And we know that all things work together for good to them that love God to them who are the called according to His purpose.**
>
> **For whom He did foreknow he also did predestinate to be conformed to the image of His Son... (Romans 8:28-29)**

"All things," both good and bad circumstances of life, are being used by God to conform you to the image of Jesus. You change as you conform to God's standards rather than the standards of the world:

> **I beseech you therefore, brethren, by the mercies of God, that ye present your bodies a living sacrifice, holy, acceptable unto God, which is your reasonable service.**
>
> **And be not conformed to this world; but be ye transformed by the renewing of your mind, that ye may prove what is that good, and acceptable, and perfect will of God. (Romans 12:1-2)**

Change only comes through the power of the Holy Spirit. Paul failed in self-improvement in

Romans chapter 7. He succeeded in change through the power of the Holy Spirit. His victory is recorded in Romans chapter 8. This is why Paul reported:

> **And He (Jesus) said unto me, My grace is sufficient for thee; for my strength is made perfect in weakness. Most gladly therefore will I rather glory in my infirmities, that the power of Christ may rest upon me.**
>
> **Therefore I take pleasure in infirmities, in reproaches, in necessities, in persecutions, in distresses for Christ's sake...for when I am weak, then am I strong. (II Corinthians 12:9-10)**

Paul recognized his weaknesses. But he did not let them discourage him. He realized God's power could change these weaknesses. If you do not identify your weaknesses and cooperate with God's process of change in your life, then you will continue to fail over and over again.

TO IDENTIFY YOUR SPIRITUAL STRENGTHS:

You are not effective in ministry by focusing on your weaknesses. You are effective by emphasizing the spiritual strengths God has given you. In the parable of the talents in Matthew 25:14-30, each servant was to focus on what he had, not what he did not have.

You identify your spiritual strengths for the purpose of ministry, not just to glory in your abilities. Every strength you have is given by the Lord, just as it was in this parable. As the servants in this parable, you are responsible to use spiritual strengths, talents, and abilities for God's glory.

Remember also that no matter how many strengths you may identify, there is always room for improvement. Paul analyzed his strengths but said he lived...

> **Not as though I had already attained, either were already perfect: but I follow after, if that I may apprehend that for which also I am apprehended of Christ Jesus.**
>
> **Brethren, I count not myself to have apprehended; but this one thing I do, forgetting those things which are behind and reaching forth unto those things which are before.**
>
> **I press toward the mark for the prize of the high calling of God in Christ Jesus.**
>
> **Let us therefore, as many as be perfect, be thus minded... (Philippians 3:12-15)**

As you identify your spiritual strengths, remember that success is not measured by what you <u>are</u>. It is measured by what you are compared to what you could be in God. A key message of the book of Job in the Bible is that even a righteous man can benefit from further experience with God.

TO IDENTIFY YOUR SPIRITUAL GIFTS:

Spiritual gifts are special abilities given by the Holy Spirit to enable the work of the ministry. The subject is of such importance that Harvestime International Institute devotes an entire manual to it.

The course *"Ministry Of The Holy Spirit"* explains spiritual gifts in detail. If you do not understand spiritual gifts or have not identified the gifts God has given you, this course will help you in this area of spiritual development.

The Bible teaches that each believer has at least one spiritual gift (I Corinthians 12:4-11) and that you are responsible to discover and develop your gifts (II Timothy 1:6).

Here is a list of the spiritual gifts:

Special Gifts: These gifts are given to be used to equip God's people for the work of the ministry.

-Apostles
-Prophets
-Evangelists
-Pastors
-Teachers

Speaking Gifts: These gifts explain God's truth:

-Prophecy
-Teaching
-Exhortation
-Word of Wisdom
-Word of Knowledge

Serving Gifts: These gifts enable God's work:

-Serving
-Helps
-Leadership
-Administration
-Giving
-Showing Mercy

-Discerning of Spirits
-Faith
-Hospitality

Sign Gifts: These gifts establish God's authority:

-Tongues
-Interpretation
-Miracles
-Healings

TO DETERMINE YOUR PLACE OF MINISTRY:

Identifying your spiritual gift is only the first step towards finding your place of ministry. After you discover your gifts, you must find where God wants you to use them in the Body of Christ.

Read I Corinthians 12 in your Bible. This chapter explains how the Church is similar to a human body. Just as each part of a human body has a special function, so does each member of the Body of Christ. To be effective, you must discover your place of ministry in the Body of Christ. To discover God's will, you must present yourself as a sacrifice to be used as He sees fit:

> **I beseech you therefore, brethren, by the mercies of God, that ye present your bodies a living sacrifice, holy, acceptable unto God, which is your reasonable service.**
>
> **And be not conformed to this world; but be ye transformed by the renewing of your mind, that ye may prove what is that good, and acceptable, and perfect will of God. (Romans 12:1-2)**

You must discover your spiritual gifts, analyze your spiritual strengths and weaknesses, and prove or examine God's will for your life. (Harvestime International Institute has a course entitled *"Knowing God's Voice"* that can assist you in this process.)

One of the sins of the flesh in Galatians 5:20 is "emulations." To emulate means to pattern yourself after someone else and imitate them. God does not want you to emulate the ministry of others. He wants you to discover your own place of service, the ministry He has planned for your life:

> **But let every man prove his own work, and then shall he have rejoicing in himself alone, and not in another. (Galatians 6:4)**

TO FORMULATE A PLAN TO FULFILL YOUR MINISTRY:

Proper analysis of your spiritual condition, strengths, weaknesses, and spiritual gifts leads to discovery of your place in the Body of Christ. It guides to revelation of your place in ministry.

But knowing your place is not enough. You must make plans to fulfill the ministry God has for you, then you must carry out the plans:

> **And that servant, which knew his Lord's will, and prepared not himself, neither did according to His will, shall be beaten with many stripes. (Luke 12:47)**

Personal analysis is not effective unless you use what you learn to fulfill your ministry.

HOW TO GATHER INFORMATION

PRAY:

Ask God to reveal your spiritual condition, strengths, weaknesses, and your place in His plan. David did this:

> **Examine me, O Lord, and prove me; try my reins and my heart. (Psalms 26:2)**

STUDY GOD'S WORD:

God's Word is the unchanging standard by which you are to examine yourself:

> **Forever, Oh Lord, thy Word is settled in Heaven. (Psalms 119:89)**

USE SPIRITUAL GIFTS:

Spiritual gifts may assist in your analysis. For example, the Lord may give you discernment regarding a weakness in your spiritual life or use someone with the gift of discernment or word of knowledge to minister to you.

SEEK COUNSEL FROM SPIRITUAL LEADERS:

Ask your spiritual leader to assist in your analysis. God gives leaders to guide you. Ask your leader to analyze your spiritual strengths and weaknesses. Ask him to pray with you regarding your specific area of ministry.

CONSULT WRITTEN RECORDS:

You may be able to use some written records in your analysis. For example, a written budget, bank statements, or checkbook records could be used in your personal financial analysis.

CONDUCT INTERVIEWS:

Consult a close family member and/or friend about your spiritual weaknesses and strengths.

OBSERVE YOURSELF:

Observe yourself in various ministry situations. Note your strengths, weaknesses, and effectiveness of service. What spiritual gifts are evident?

Observe yourself in various real life situations. How do you react under pressure or stress? Is spiritual fruit evident in your life?

PERSONAL ENVIRONMENTAL ANALYSIS

Use the following outline to do your personal analysis:

ANALYZE YOUR RELATIONSHIPS

God's Word and all Christian life is based on relationships. Because of this, it is important to analyze your relationships with God and others:

> **Jesus said unto him, Thou shalt love the Lord thy God with all thy heart, and with all thy soul, and with all thy mind.**
>
> **This is the first and great commandment.**
>
> **And the second is like unto it, Thou shalt love thy neighbor as thyself.**
>
> **On these two commandments hang all the law and the prophets. (Matthew 22:37-40)**

Analyze your relationships with the following:

GOD:

-Are you a born-again believer?

-Have you been baptized in water?

-Are you walking in fellowship with God? Is there any known sin in your life with which you need to deal?

-Have you been baptized with the Holy Spirit?

-Do you read your Bible regularly?

-How could you improve your Bible study habits?

-Do you pray regularly?

-How could you improve your prayer life?

-Are you an active part of a local Church fellowship?

-Does your life evidence the fruit of the Holy Spirit? (For a listing of these see Galatians 5:22-23.)

-In what areas of your spiritual relationship with God do you feel you are immature and need development?

YOUR MATE:

-Do you have a good relationship with your mate? (Do you get along well, communicate, spend time together, etc.)

-Do you pray, study God's Word, and attend Church together? If not, what could you do to include these practices in your relationship?

-What are some of the strengths of your relationship with your mate?

-What are some of the weaknesses in your relationship with your mate?

-How could you improve these weaknesses?

-Does your relationship with your mate hinder your ministry?

-What could you do to make your relationship with your mate a benefit rather than a hindrance to God's work?

YOUR EMPLOYER:

-Do you have a good relationship with your employer?

-If not, how could you improve it?

-Do you give him respect and honor?

-Do you do all things on your job as unto God? (See I Corinthians 10:31.)

YOUR CHILDREN:

-Do you have a good relationship with your children?

-How could you improve your relationship with your children?

-Are your children well disciplined?

-In what areas are they not well disciplined?

-What could you do to improve discipline in these areas?

-Have your children accepted Jesus as Savior?

-What could you do to help your children make this decision?

-Analyze the spiritual growth of your children who are believers. How can you assist them in further spiritual development?

-If you are a leader in the Church, do you have your household under control as indicated in Titus 1:6 and I Timothy 3:4-5? If not, what changes need to be made to meet this standard?

OTHER FAMILY MEMBERS:

-Do you have difficulty in relationships with any family member? If so, who and why?

-What could you do to improve this relationship?

OTHERS IN THE BODY OF CHRIST:

-Do you have difficulty in relationships with other members of the Body of Christ? If so, with whom do you have difficulty?

-What can you do to restore your relationship with this Christian brother or sister? (Scriptural methods for restoring fellowship with another believer are given in Matthew 18:15-17.)

UNBELIEVERS:

-Are you in contact with unbelievers or have you surrounded yourself only with Christians? (You cannot win the lost if you have no contact with them.)

-What steps could you take to win to the Lord unbelievers with whom you have contact?

-How could you establish more contact with unbelievers for the purpose of sharing the Gospel?
-Do you have any relationships with believers that are unscriptural and hurting your own relationship with God? If so, what can you do to correct these situations?

ANALYZE YOUR PHYSICAL CONDITION

-The Bible says your body is the temple of God's Holy Spirit. Since your body is God's dwelling place, you are to properly care for it (I Corinthians 3:16-17).

-Your physical condition affects you spiritually because man is body, soul, and spirit. When you are not well cared for physically, you can become easily discouraged spiritually. You do not have the energy and strength to do the work of God.

-Are you seriously overweight?

-Are you seriously underweight?

-Do you use harmful substances such as alcohol, cigarettes, or drugs?

-Do you try to eat as well as your budget and/or environment permits?

-Do you exercise regularly?

-Do you have any physical conditions that hinder your work for the Lord. If so, what can you do about them? (Some things you can improve by diet and exercise. Others God can heal.)

ANALYZE YOUR EDUCATIONAL LEVEL

-What is your present educational level?

-What can you do to improve your education to benefit the work of the ministry? (Studying Harvestime International Institute courses is one way! Do not be discouraged if you have not had the opportunity to receive much education. Jesus chose common men like fishermen and used them mightily in ministry. If you do have opportunity to improve your education and you feel it is God's plan for you, we encourage you to do so.)

ANALYZE YOUR FINANCES

-Do you tithe regularly? (A tithe is 10% of all money you make given to God. This is a Biblical principle.)

-Do you give offerings in addition to your tithe?

-What was the total amount of your giving to the Church last year?

-How much did you give to missions?

-What adjustments could you make in your standard of living which would result in your having more to give to the work of the Lord?

-Are you in debt? If so, what can you do to eliminate these debts?

-Do you have a good testimony regarding how you handle your finances, provide for your family, pay bills, etc?

-If not, how could you improve in this area?

-Are you overly concerned with gaining wealth and overly attached to material things? Do these things occupy most of your time and energy?

ANALYZE YOUR SPIRITUAL ABILITIES

-Do you make wise use of your time for spiritual purposes and the work of the ministry?

-What activities could you eliminate which would give you more time to minister?

-Do you know the specific ministry the Lord has for you in the Body of Christ?

-Are you fulfilling it now? If not, why not?

-Do you know what spiritual gifts you have been given? If not, take the following steps:

STEP ONE: Be born again.

You must be born again. Spiritual gifts come through the new birth just as natural talents come through natural physical birth. If you are never born in the natural world you cannot have natural talents. If you are not born again in the spiritual world you cannot be given spiritual gifts:

Repent, and be baptized every one of you in the name of Jesus Christ for the remission of sins, and ye shall receive the gift of the Holy Ghost. (Acts 2:38)

STEP TWO: Receive the baptism of the Holy Spirit.

Guidelines for how to receive the baptism of the Holy Spirit are given in the Harvestime International Institute course, *"Ministry of the Holy Spirit."*

STEP THREE: Know the spiritual gifts.

You must know what the spiritual gifts are. If you do not know, you will not be able to recognize the spiritual gift God gives you.

STEP FOUR: Seek a spiritual gift.

But covet (desire, seek) earnestly the best gifts. (I Corinthians 12:31)

We each already have at least one gift, but we can seek additional gifts. This verse implies you can receive a gift which you do not now possess.

STEP FIVE: Laying on of hands.

Have your spiritual leader lay hands on you and pray for God to reveal to you your spiritual gift:

Neglect not the gift that is in thee, which was given thee by prophecy, with the laying on of the hands of the presbytery. (I Timothy 4:14)

As a result of Steps One through Five God may reveal your spiritual gift. If this does not happen, proceed with the following steps.

STEP SIX: Analyze your spiritual interests.

What gifts would you like to see developed in your life? What gifts bring you the most joy in thinking about them or using them. Do you enjoy teaching? Do you enjoy being hospitable and having people into your home? Are you often moved to give large sums of money to God's Work?

STEP SEVEN: Analysis by a Christian leader.

Have a Christian leader analyze your spiritual abilities. What have they observed about your Christian service? Do you appear to them to be a good administrator, teacher, etc? In their opinion, in what gifts or ministries have you been effective?

STEP EIGHT: Analyze your past Christian service.

In what areas of gifts/ministries did you experience great joy as you served? In what areas were you most effective?

STEP NINE: Identify gifts you think you might have.

Make a list of the gifts you think you might have. This list is based on:

-The knowledge of gifts you have obtained through study.

-What God has revealed to you through prayer.

-What a key Christian leader has observed in your life.

-Analysis of your effectiveness in areas of ministry in which you have previously served.

(The Harvestime International Institute course entitled *"Ministry Of The Holy Spirit"* will also assist you in identifying your gifts.)

STEP TEN: Look for spiritual needs around you.

What needs are not being met in your neighborhood, community, or church? Is there a need for teachers, evangelists, administrators, etc Make a list of these needs.

STEP ELEVEN: Fill a spiritual need.

Compare the list of spiritual needs to the list of gifts you believe God has given you. Recognizing a need does not necessarily mean you have the gift to meet that need. But you can determine where a spiritual need corresponds with the gift you believe God has given you.

Make a commitment to fill this need. For example, if there is a need for teachers in your church and you believe you have the gift of teaching, volunteer to meet this need.

STEP TWELVE: Evaluate your ministry.

After serving awhile in this area with your gift, evaluate your ministry. Are you effective in using this gift? Does it bring great joy to you? Do you see results? Does your pastor or spiritual leader see results? If so, God has led you to discovery of your spiritual gift.

If the area you selected does not fit your spiritual capabilities and you are ineffective, review your list of possible gifts and ask God to show you another area in which to minister.

APPLY THE RESULTS OF YOUR ANALYSIS

The analysis you have completed will be of no benefit unless you apply it to your life and ministry. To "apply" something means to use it effectively, to actually do something with it.

Anytime you want to accomplish something....to do any work for God...you must have a plan. God is a God of planning. He planned this world and carried out His plan in creation. From the foundation of the world, God made a plan for salvation for sinful mankind. When it was the proper time, He fulfilled that plan by sending Jesus into the world to die for the sins of all people.

Whenever Jesus was going to work in the lives of men and women, He called upon them to prepare. He told blind men to go wash in pools of water. He commanded disciples to bring him the loaves and fishes. He told the lame to take up their bed.

You must cooperate with the work of God in your life. To be sure you do this, you must plan to do it. When you apply the results of your personal analysis, you plan to use what you have learned about yourself for God's glory.

Use the following guide to apply the results of your analysis. Remember to pray about each section. Ask God to reveal His plan to you, then write each plan in the spaces provided.

1. DETERMINE YOUR SPIRITUAL CONDITION:

Review your analysis and write a paragraph summarizing your present spiritual condition. Include in this summary the ways your spiritual condition is affected by your relationships, your physical condition, your finances, education, strengths, and weaknesses.

Make a plan: What can you do to cooperate with God in improving your spiritual condition, increasing in growth and maturity?

I will:

____

____

2. IDENTIFY YOUR SPIRITUAL WEAKNESSES:

Review your analysis and list your spiritual weaknesses. Beside each weakness write your plans for dealing with that weakness:

WEAKNESS	MY PLAN TO DEAL WITH IT I WILL...
____________________	____________________
____________________	____________________

3. IDENTIFY YOUR SPIRITUAL STRENGTHS:

Review your analysis and list your spiritual strengths. Beside each strength write your plan for using it for God's glory:

STRENGTH	MY PLAN TO USE IT FOR GOD'S GLORY I WILL...
______________________	______________________
______________________	______________________

4. IDENTIFY YOUR SPIRITUAL GIFTS:

Review the results of your spiritual gifts analysis. List the spiritual gifts you believe you have. Beside each gift write your plan to use that gift for God's glory:

SPIRITUAL GIFT	HOW I WILL USE THIS GIFT I WILL...
______________________	______________________
______________________	______________________

5. DISCOVER YOUR PLACE OF MINISTRY:

Review your entire personal analysis. Prayerfully consider and complete this statement:

I believe my place of ministry in the Body of Christ is:

____

____

6. FORMULATE A PLAN TO FULFILL YOUR MINISTRY:

How will you begin to fulfill the ministry you believe God has given you? Study Chapter Eleven to help formulate a plan.

____

____

SELF-TEST

1. Write the Key Verse from memory.

____

____

2. Define the term "personal environmental analysis."

____

____

3. What does a personal environmental analysis include?

____

____

4. List six purposes of a personal environmental analysis.

____

____

____

____

____

____

(Answers to tests are provided at the conclusion of the final chapter in this manual.)

FOR FURTHER STUDY

1. Read the story of Samson in Judges 13-16. Here was a man who could have benefitted from a personal analysis. He did not realize his own true spiritual condition when he went out to battle the Philistines. If Samson had done a personal analysis...

   -What were some of the spiritual problems his analysis might have revealed?

   -Could he have made changes that would have affected the destiny of his life?

   -What changes could he have made?

2. Read Psalms 51 where David is analyzing his personal spiritual condition after his sin with Bathsheba. From your study of this chapter, answer these questions:

   -What was David's spiritual condition when he wrote this chapter?

   -What did David do about his sin?

   -What did David ask God to do?

3. Read the story of Belshazar in Daniel 5. Here was another man who could have benefitted from personal analysis.

   -What was his spiritual condition? (See verses 22-23.)

   -What did God do about it? (See verses 25-28.)

   -If Belshazar had been aware of his own personal spiritual condition, what changes could he have made in his life that might have avoided God's judgment?

4. Read the story of Nebuchadnezzar in Daniel chapter 4. Here was a man who analyzed his environment incorrectly and suffered because of it.

   -Compare his environmental analysis in verses 29-30 to that in verses 34-35. How do they differ?

   -What happened to make him aware of his true spiritual condition? (See verses 31-33.)

   -What happened when Nebuchadnezzar analyzed his personal situation properly and acknowledged all he had was from God? (See verses 34-37.)

5. In Luke 12:16-21 read the story of the rich young man who left spiritual matters out of his personal analysis. His personal analysis was based on material things of this world rather than things with eternal value.

6. Read Luke 18:9-14. Compare the personal analysis of the Pharisee to that of the sinner. Which analysis was honest?

7. The book of Judges in the Bible gives the record of several kings who ruled God's people. Some of them were good kings, others were bad. Read through the book of Judges and do a personal analysis of each king. What were there strengths and weaknesses? How could they have benefitted from personal analysis? What changes could they have made which would have affected their destiny?

8. Read Numbers 12. Miriam and Aaron did not properly analyze their own spiritual position in God. What happened as a result?

9. Korah and his followers made a similar mistake in Numbers 16. What happened as a result of their faulty analysis?

10. Read I Kings 19. How did Elijah analyze his environment at this time? How did God correct his analysis?

CHAPTER SIX

ANALYZING A PEOPLE GROUP

OBJECTIVES:

Upon completion of this chapter you will be able to:

- Write the Key Verse from memory.
- Define the term "people group."
- Explain what a people group analysis includes.
- Identify purposes of a people group analysis.
- Analyze a people group.
- Use your analysis to spread the Gospel to this group.

KEY VERSE:

> **Neither is there any creature that is not manifest in His sight: but all things are naked and opened unto the eyes of Him with whom we have to do. (Hebrews 4:13)**

INTRODUCTION

In the last chapter you analyzed yourself in preparation for ministry. This chapter begins a series of lessons which explain how to analyze different environments in which you will minister.

In this chapter you will learn how to analyze a people group. You will learn in following chapters how to analyze areas, nations, and regions of the world which are made up of many different people groups.

A PEOPLE GROUP

The people of this world are divided by a number of barriers which may affect the spread of the Gospel. There are geographic barriers such as mountains, oceans, and valleys. There are also political boundaries that separate countries.

But people are also separated by cultural barriers. In general, culture refers to the differing ways people live in our world. Culture includes such things as language, beliefs, values, social structure, customs, etc.

These cultural barriers have resulted in the development of various "people groups" throughout our world. A people group is any group of people where marriage and intimate life generally takes place only within the society. For example, a people group is a specific tribe, caste, clan, or lineage.

When Jesus told His disciples to preach the Gospel to all nations, the word He used for "nations" was the Greek word "ethnos." A nation is made up of many groups of "ethnos" or "people groups."

It is not enough just to reach a nation in general with the Gospel. You must be sure that every people group within that nation hears the good news. When this is done, the whole nation will be evangelized. Jesus commanded that the Gospel be preached to all peoples. If you are to fulfill this command, you must analyze people groups to determine how they may be reached, make a plan to reach them, and then put the plan into action.

ANALYSIS OF A PEOPLE GROUP

When you analyze a people group, you study a specific group of individuals who are part of the same tribe, clan, or lineage. You study them socially, culturally, politically, and spiritually. You also analyze external forces from outside their group which affect their acceptance of the Gospel.

THE PURPOSE OF ANALYZING A PEOPLE GROUP

The purpose of your study is to reach a people group with the Gospel. It is research for the purpose of evangelism, not just for the sake of research. Be careful that you do not spend all of your time studying and never get around to accomplishing the purpose which originally motivated your study!

Your study of the culture of a people group will help you determine how to best reach them. You can present the Gospel in a way that emphasizes what you have learned about their culture.

For example, when Paul ministered to Jews, He emphasized the Jewish lineage of Christ and presented Him as Israel's Messiah. When he ministered to Gentiles (others nations), he stressed different aspects of the Gospel.

Other examples are the books of Matthew, Mark, Luke, and John. They all emphasize different aspects of the Gospel which relate to the culture of the specific groups for which they were written.

When you use the culture of people as a bridge to present the Gospel it is more readily accepted. Understanding culture can give you spiritual keys which will unlock the door to a particular people. You analyze a people group to answer these questions:

WHAT PEOPLE ARE TO BE REACHED?

Who are you really trying to reach with the Gospel? Where is this group located?

WHAT ARE THEY LIKE?

What do you need to know about their customs, problems, social structure, spiritual condition, etc., in order to reach them?

WHO SHOULD REACH THEM?

Will these people be reached easiest through a man or woman, through one of their own people or an outsider? What qualifications will the person need in order to reach them? God has special messengers for reaching specific people groups. Paul was used by God to reach the Gentiles. Peter was used most effectively to reach the Jews.

HOW SHOULD THEY BE REACHED?

What methods should be used? Which will be most effective? The answers to these questions will help you determine your role in reaching these people and formulate a plan to reach them with the Gospel.

HOW TO GATHER INFORMATION

You can gather information for your analysis by:

PRAYER:

Ask God to reveal what you need to know about this group and help you reach them with the Gospel. God has divine knowledge of all peoples (Hebrews 4:13).

STUDY OF WRITTEN MATERIALS:

Study books and written materials about the people group. A list of possible sources of information is included in the Appendix of this manual.

OBSERVATION:

Personally observe the group yourself.

INTERVIEW:

Ask questions of people who are leaders and members of the group you are studying. If there are believers already ministering to these people, interview them to discover what they have learned about the people. (Sample interviews are included in the Appendix of this course.)

Use the following outline to analyze a people group. You may not be able to answer all of the questions about a specific group of people, but answer those for which you can obtain information.

WHAT PEOPLE ARE TO BE REACHED?

1. Who are the people you are trying to reach?

2. What makes them a "people group"? They should have several of the following factors in common. Check those which apply to the group you are studying:

   _____Same language.
   _____Common political loyalty.
   _____Racial or ethnic similarity.
   _____Shared religious beliefs.
   _____Common kinship ties.
   _____Similar educational level.
   _____Common residential area.
   _____Similar social class or caste.
   _____Shared culture (accepted ways of doing things).
   _____Shared special interest.
   _____Shared economic level (i.e., all poor, middle class or rich).
   _____Unique health situation.
   _____Different legal status (i.e., all prisoners, refugees, illegal residents, etc.).
   _____Similar age.
   _____Common significant problems.
   _____Experience discrimination from others because they are part of this group.

   What other common factors can identify them as a distinct group?

3. How many people are there in this group? Make a general estimate of the population if exact figures are not available. This will help determine the number of workers needed to reach them with the Gospel.

GEOGRAPHICAL ANALYSIS

1. In what specific area do the majority of these people live?

2. In what other geographic areas are they found?

   (Note: In answering these questions identify the people specifically, not just as "all the people in India." Which *specific* people? For example, in India you would identify a specific caste in a specific village. The methods you use to reach each caste and each village may differ.)

3. Obtain or draw a map of the country and mark the location of the specific group you want to reach.

4. Obtain or draw a map of the local area where these people live. Begin to pray over these maps that God will reach the people in the areas you have marked. Also use these maps to answer the following questions:

5. What geographic factors will hinder the spread of the Gospel? (For example, is the area where they live difficult to reach?)
6. What geographic factors might be used to help the spread of the Gospel? (For example, are there natural routes of trade or travel along which churches could be planted?)

7. Does the location of this group make certain times of the year best for entering and reaching the people? (For example, traveling into a remote area during the heavy rainy season might not be as practical as entering during the dry season, unless God would specifically lead you to do so. The roads may be impassable.)

8. Obtain information on the following factors which may hinder or assist in reaching these people in their present location:

   -Distance to nearest major city:
   -Distance to capitol of country:
   -Distance to established airport:
   -Distance to usable airstrip or water landing:
   -Do they have telephone or Internet service?
   -Do they have telegraph service?
   -Distance to nearest rail station:
   -Distance to nearest usable automobile road:
   -Distance to nearest navigable river or sea ("Navigable" means that you can use a boat to travel):

WHAT ARE THEY LIKE?

SOCIAL ANALYSIS:

1. What is the basic social unit of this people group? For example, is it one husband/one wife and their children or are there several wives? Do people live separately in the basic family unit of a father, mother, and children? Or does the extended family live together (aunts, uncles, and other relatives all in the same household)?

2. Who is the leader in the home? Is it the father, mother, mother-in-law, other? You can use family relationships as bridges over which the Gospel can spread. For example, if in a certain culture the father rules the home, he is your first target to reach with the Gospel. His family will most likely follow after he accepts the message.

3. Who are the community leaders? Who is the tribal chief, head of the clan or caste, leader in the village or town where these people live? If you reach them with the Gospel they can lead others to acceptance.

4. Who makes the decisions in this group and how are decisions made? You will need to know who makes decision and how they are made if you must obtain permission to enter the village or tribe to share the Gospel.

5. What is their community involvement? With what are they concerned as a group? You can emphasize these concerns in your presentation of the Gospel. Relate the message to meet these needs.

6. How open are they to "outsiders" (people who are not part of their own people group)? Knowing this will prepare you for the level of acceptance you will experience when you first enter this group to share the Gospel.

7. What is their social level? Are they upper class, middle class, lower class? If the people are lower class with great physical or material needs, you may want to use a medical ministry or feeding program as a method to enter their group to share the Gospel with them.

8. What occupations are represented? How do these people make their living? Think about how you can use their occupations to gain entry to their group. For example, you may want to start a Bible study group at their work place or you might be able to share different methods of agriculture with farmers, etc.

9. Who controls the financial and material resources of the group and/or within each family unit (credit, markets, lands, jobs)? You may need their assistance to obtain property for a church and finances for spreading the Gospel.

CULTURAL ANALYSIS:

Study cultural practices that may affect their acceptance of the Gospel. Here are some to consider:

1. Beliefs And Values:

   a. What cultural values, beliefs, or practices of the people might <u>help</u> them understand the Gospel? (For example, if the people have strong positive feelings about the importance of the family, the Gospel's emphasis on the family would be a bridge to them.)

   b. What cultural values, beliefs, or practices of the people might <u>hinder</u> their understanding of the Gospel?

2. Cultural Practices:

Are there some common cultural practices that are clearly prohibited by Scripture? If so, what are they? Practices are not wrong just because they are different from the way you do things. Practices are wrong when they conflict with God's Word. Sinful practices will have to be dealt with as you reach these people with the Gospel. Other cultural practices are not necessary to change. People do not have to accept your culture to accept the Gospel.

3. Needs:

   a. What are the general concerns, attitudes, motives, feelings, commitments of the people? Which of the following do they consider as important needs? Do they have...

   _____Inadequate food.
   _____Poor health and medical services.
   _____Poor education.
   _____Improper water and/or poor supplies of it.
   _____Inadequate financial resources.
   _____Inadequate geographic resources.
   _____Poor transportation.
   _____Poor systems of communication.
   _____Inadequate housing.
   _____Inability to cope with change.
   _____Political oppression.
   _____Other:___________________________________

Your presentation of the Gospel and perhaps your methods of ministry in their culture can relate to these needs.

b. What needs appear to be well met? What needs do they NOT have? It is important to know the needs being met as well as those not being met. For example, you would not want to use a feeding program as a method to enter a group to share the Gospel if there are no hungry people and everyone is well fed.

4. Education:

What is their educational level? What is the literacy level? How many of the people read and write their own language? It is important to know this because you would not want to design a Gospel outreach based on literature if the people do not read and write.

5. Communication:

a. Which language (or languages) is most suitable for presentation of the Gospel? You want to reach the largest number of people in a group first, so select as first choice the language which most of the people speak.

The main language spoken by people in this culture is_______________.
Other languages needed later to reach the remainder of the people:____________.

b. What are the main methods of informal, social communication? Does communication flow best through tribal or community leaders? Along family lines? Every group has a communication network by which important news is spread through various levels.

c. Can these people be reached through:

_____Newspapers (Do they accept Christian advertisements?)
_____Printed literature (Can the people to be reached read?)
_____Radio (Do the stations permit Christian broadcasting?)
_____Television (Do the stations permit Christian broadcasting?)
_____Tape recordings (Do the people have access to tape recorders?)
_____Movies, video tapes (Is the equipment available in this area?)
_____Telephone or Internet (Do the people have telephones or Internet?)

d. What is the main method of communication?

EXTERNAL INFLUENCES:

1. What changes are taking place in this group as a result of the political system or outside forces of modernization?

2. Can any of these changes be used to help spread the Gospel?

3. Are there government restrictions that in any way prohibit the evangelization of this people? If so, what are they?

4. What methods could be used to reach this group that would not conflict with these regulations?

5. What organized groups or agencies may oppose the spread of the Gospel among this people group? Why?

   Never let restrictions or opposition prevent you from ministering to a group to which you feel God has called you. If you are not allowed to reach them in certain ways, use other methods.

POLITICAL ANALYSIS:

1. Briefly describe the political system of these people.

2. What political groups among these people might hinder the spread of the Gospel?

3. What political groups among these people might help in the spread of the Gospel?

4. What government regulations might hinder your ministry with these people?

5. What government regulations might assist in your ministry to these people?

6. Who are the key political leaders among these people? You might try to make their acquaintance and try to gain their favor for your ministry to these people.

SPIRITUAL ANALYSIS:

1. What religious groups are found among this people?

2. How many of the people follow each religion?

3. What makes them unhappy with their current religion? For example, magical practices that do not work for disease, etc. You can focus the Gospel message to target these dissatisfactions.

4. Is the practice of their religion increasing or decreasing?

5. Does their religion have any written scriptures, (i.e., the Muslims with the Koran, etc.)? If so, how can you use them as a bridge to present the Gospel message? (This may be possible with some, impossible with others.)

6. What is the attitude of this people towards religious change of any kind? The answer to this question will prepare you for religious opposition you might be facing.

7. Answer the following questions about their knowledge of the Christian faith. These answers will help you determine where you must begin in your presentation of the Gospel:

What is the attitude of this people towards...

God:

____They believe a God exists.
____They deny the existence of a God.
____They know about the one true God as revealed in the Bible.
____They do not know about the one true God.
____They know about the one true God but are indifferent toward Him.
____They know about the one true God and are hostile toward Him.

Jesus Christ:

____They believe Jesus exists.
____They deny the existence of Jesus.
____They know that Jesus died for their sins.
____They do not know Jesus died for their sins.
____They know about Jesus but are indifferent toward Him.
____They know about Jesus and are hostile toward Him.

The Holy Spirit:

____They believe the Holy Spirit exists.
____They deny the existence of the Holy Spirit.
____They never heard about the Holy Spirit.
____They do not understand the ministry of the Holy Spirit.
____They understand the ministry of the Holy Spirit but are indifferent toward Him.
____They understand the ministry of the Holy Spirit but are hostile toward Him.

The Gospel:

____They have no awareness of the Gospel.
____They have some knowledge of the Gospel.
____They understand the basic truths of the Gospel.
____They understand that it is necessary to respond personally to the Gospel.
____They recognize their need of the Gospel.
____They have accepted the Gospel.

The Bible:

____They never heard about the Bible.
____They know the Bible exists.
____They believe it is the Word of God.
____They are hostile toward the Word of God.
____They are indifferent toward the Bible.

What is the status of Scripture translation in the language that is most suitable for reaching these people? What portions of the Scripture remain to be translated in their own language?

New Testament:

Old Testament:

Christians:

How aware are they of Christians?

____They never heard of a Christian.
____They have heard, but never met one.
____Have met some Christians.
____Have true believers living among them.

8. If there are Christians in this group, answer these questions:

-How many people in this group claim to be Christians?

-How many of these have been incorporated into a fellowship of believers (a church)?

-How many are active in spreading the Gospel.

-How many were Christians but have turned back from following Christ?

-What were the reasons that seemed to draw them away from God?

-Who are the key Christians among this group?

-Are there any "tent maker" believers living among these people? (Believers working at a secular occupation among these people while actively sharing the Gospel.)

9. In general, do you believe this is a time for sowing the Gospel among these people or a time for reaping? Usually, more workers are needed when an area is spiritually ripe for harvest.

10. What Christian churches or missions have been attempted in this group and failed? What do you believe were the reasons for their failure? You can learn from the failures of others and not repeat the same mistakes.

11. What Christian churches or missions are now active within the area where the majority of this people group lives? Obtain their names and addresses.

12. Which of these are effective and why are they effective? You can learn from the successes of these groups. You will also want to cooperate with them in spreading the Gospel to these people. How can you cooperate with them to reach this people group?

13. If there is a Christian witness present, what language or languages is being used? You may be led to reach people in this group who speak a different language and are not currently being reached by the present Christian witness.

14. What types of Christian ministry are presently being done among these people? For example, if there is already an effective feeding program for the hungry being used as a Gospel witness, do not duplicate efforts. Your goal is to reach the unreached and/or expand efforts to reach these people, not duplicate efforts or compete with existing ministries.

15. What is the attitude of the people toward these existing Christian missions?

_____Hostile
_____Indifferent
_____Friendly

16. If there are true Gospel churches among this group, answer these questions:

-How many true Gospel churches have been established among these people?
-How fast are they growing?
-What difficulties are they experiencing and why?
-Which ones are successful and why?

-Is there a need to establish more churches?

17. If there is no existing Christian witness among the people you are studying, has another Christian agency had experience reaching a similar group? You could draw on their experience in a similar group to reach the group you are studying.

18. Can you determine what principalities and powers of Satan are active in this people group? The Bible teaches that there are certain spiritual forces active among certain groups and over certain geographic areas. (See Daniel 10. Note especially verse 13. Also see Ephesians 6:10-17.)

    Prayer will usually reveal what principalities and powers are in operation. Sometimes observation and study of the history of a people, area, nation, or region will also assist in identifying the spiritual forces in operation.

    For example, in one place in the United States there are many homosexuals. It is quite easy to identify by observation the ruling spirit of homosexuality in this people group and their city. Another city in the United States developed because of a rush of people to search the area for gold. The spirit of greed is still ruling that city today.
    When you identify these spiritual forces, you must bind them in prayer and loose the power of the Holy Spirit to work among these people. The Harvestime International Institute course entitled *"Spiritual Strategies: A Manual Of Spiritual Warfare"* provides further study on this subject.

WHO SHOULD REACH THEM?

Who should reach these people? Who can speak their language? Who would be accepted by their culture and social structure? Use the information you gathered on their culture and social structure to help you determine this. Also use the information on Christian agencies that have tried and been successful or tried and failed. Who they used may have affected their success or failure.

Whoever goes to share the Gospel should...

-Be a believer with evidence of this in their lifestyle, attitudes, and behavior.

-Have good personal prayer and Bible study habits.

-Have spiritual gifts properly suited to the methods to be used to reach the people.

-Be available: Ability is no good without availability. Is the person willing and ready to go?

-Be properly fitted into the body (willing to cooperate with other members for the goal of reaching these people with the Gospel).

-Be called of God to minister to these specific people. God calls people to specific groups, just as Paul was called to reach the Gentiles and Peter the Jews.

-Be adequately prepared physically, spiritually, and educationally to reach this specific group. Remember that Gideon had to trim down his army to those few who were particularly equipped.

HOW SHOULD THEY BE REACHED?

How should these people be reached? What methods should be used? To answer this question, use what you learned about their culture and social structure. Focus methods on what will meet their needs. Use the best methods of communication. Be sure your methods are Scriptural. Do not use any method that conflicts with Biblical principles. Here is a list of various methods that have been used as part of evangelistic strategy in different areas of the world. Which of these might be effective in reaching this people group?

_____Radio broadcasts.
_____Television programs.
_____Christian literature (Books, magazines, tracts).
_____Audio cassettes.
_____Video cassettes.
_____Films.
_____Bible study classes (In the home, church, or a community meeting place).
_____Correspondence courses.
_____Bible schools and colleges.
_____Training for Christian laymen.
_____Secular newspapers/magazines.
_____Witnessing from house to house.
_____Musical concerts.
_____Special evangelistic services in the church.
_____Open air meetings (Crusades, streets, parks).
_____Christian schools for children providing elementary and secondary education.
_____Health care.
_____Emergency relief (Care given in time of famines, disasters, etc.).
_____Employment and job skills assistance (where unemployment is a major problem).
_____Food distribution.
_____Agricultural assistance.
_____Christian art and drama.
_____Specialized outreach teams for youth, children, men, women, etc.
_____Literacy work (Helping people learn to read and write).
_____Child care assistance for working parents.
_____Jail ministry.

_____Ministry to special groups (Physically or mentally handicapped, drug addicts, alcoholics, unwed mothers, elderly, etc.).

After you have determined the methods by prayer and analysis, ask this question: What opposition may be raised against these methods by either the people group or external forces? How can you deal with this opposition?

FORMULATE A PLAN

The analysis you have completed is of no benefit unless you use what you learned about the people you studied. These people will not be reached by study alone. You must make a plan to reach them with the Gospel. In the last chapter you made a plan to fulfill your own personal ministry. In this and following chapters you will make a plan to reach people, geographic areas, nations, and regions of the world. Chapter Eleven, "A Strategy For Planning," will help you do this. (Credit note: Some of the questions were drawn from *"That Everyone May Hear"* by Edward R. Dayton.)

SELF-TEST

1. Write the Key Verse from memory.

____

____

2. Define the term "people group."

____

____

3. What does a people group analysis include?

____

____

4. What are the purposes of a people group analysis?

____

____

5. As part of your study of this lesson, did you actually complete the analysis of a people group?

____

6. Have you made a plan to use what you learned to spread the Gospel to this group?

____

(Answers to tests are provided at the conclusion of the final chapter in this manual.)

FOR FURTHER STUDY

Study Hosea chapter seven. This chapter is God's spiritual analysis of the people groups known as the Ephraimites and the Samaritans which were part of the nation of Israel. These people did not realize their true spiritual condition (verse 9). "Grey hairs," when applied spiritually, is a sign of declining spiritual strength and approaching spiritual death.

In the natural world it is one thing to have gray hairs, it is another thing to have them and not know it. It is contrary to nature for a person to have gray hair and not know it. God's problem with His people is not that they have gray hairs, but that they have them and do not know it. Having gray hairs and not knowing it is an example of an unexamined life. Verse 2 records that they "considered not in their hearts." In their own hearts, they did not recognize the signs of declining spiritual strength and approaching spiritual death (gray hairs).

Here are some things that caused their declining spiritual strength. Examine your own heart as you study this list:

1. A DECEITFUL APPEARANCE:

God used an oven to illustrate this in verses 6-7. The fire in the oven appeared to be out, but in reality was smoldering. Do you know people whose outward appearance is good? They act like a Christian, talk like a Christian, do good works, but inside the fires are smoldering. There is hurt, resentment, anger, criticism, envy, greed, unforgiveness. God opens the oven door to reveal the problem. He allows circumstances to come into our lives to enable us to discover we do not have the victory we thought we had.

2. PARTIAL OBEDIENCE:

God uses a half-baked cake to illustrate this in verse 8. They had done some things God commanded but left others undone. Partial obedience is total disobedience. A women who is 85% faithful to her husband is actually 100% unfaithful to Him.

3. UNSTABLE LOYALTY:

This is illustrated by the example of a silly dove in verse 11. These people followed any voice that promised help, i.e., Egypt, Assyria, etc. Looking here and there will not solve our problems. The solution is not in Egypt or Assyria. It is in the Lord Jesus Christ.

4. UNRELIABLE SERVICE:

This is illustrated by a faulty bow in verse 16. God compares these people to a slack bowstring that has lost its effectiveness. If we are unreliable in our Christian service we continually fall short of the goal. We are never driven to the target. We never reach the purpose God has for our lives.

God wants to correct these difficulties in our lives also. He said in verse l, "They return...but not to me." God wants to heal, revive, bless, and use us but we are reluctant to meet the requirement which is that we must return to Him. We sometimes turn, but just enough to solve our problem. We do not return to God. It is not enough to just turn toward God, we must return to Him in total surrender.

CHAPTER SEVEN

ANALYZING AN AREA

OBJECTIVES:

Upon completion of this chapter you will be able to:

- Write the Key Verse from memory.
- Define the term "area" as it is used in this study.
- Explain what an area analysis includes.
- Identify purposes of an area analysis.
- Analyze an area.
- Use your analysis to spread the Gospel within this area.

KEY VERSE:

> **Arise, walk through the land in the length of it and in the breadth of it; for I will give it unto thee. (Genesis 13:17)**

INTRODUCTION

In the last chapter you analyzed a people group. In this lesson you will analyze a specific area for the purpose of reaching it with the Gospel.

AN AREA

For purposes of this study, an area is defined as a specific portion of a country determined by either geographical, linguistic, cultural, or political lines.

A geographic area, for example, could be a valley surrounded by mountains which separates it from the rest of the nation. The boundaries of a geographic area are set by natural elements.

A linguistic area is all the territory within a nation where the people speak the same language.

A cultural area is a territory within a country where all the people groups are similar. For example, they may all be tribal groups or all followers of a certain religion.

A political area is a township, city, county, province, or state within a nation. Its boundaries are set by the government.

ANALYSIS OF AN AREA

When you do an area analysis, you analyze in detail a specific area within a nation. You determine the political, geographic, linguistic, and cultural factors that make it a separate area. You analyze its spiritual condition. You also analyze external forces from outside the area which might affect acceptance of the Gospel.

THE PURPOSE OF ANALYZING AN AREA

The purpose of your study is to reach a specific area of a country with the Gospel. It is research for the purpose of evangelism, not just for the sake of research. Be careful that you do not spend all of your time studying and never get around to accomplishing the purpose which originally motivated your study! The things which unite an area and make it distinct can provide you with a general strategy to reach the entire area with the Gospel.

You analyze an area to answer these questions:

WHAT PEOPLE GROUPS ARE WITHIN THIS AREA?

Who are they, how many, and where are they located?

WHAT DO THESE GROUPS HAVE IN COMMON?

What unites these people groups as part of a distinct area?

WHAT IS THIS AREA LIKE?

What do you need to know about this area in order to reach the people who live there with the Gospel?

WHO SHOULD REACH THIS AREA?

What are the necessary qualifications for workers to spread the Gospel in this area?

HOW SHOULD THIS AREA BE REACHED?

What common methods can be used in the whole area? Which will be most effective? The answers to these questions will help you determine your role in reaching an area and assist you in formulating a plan to reach it with the Gospel.

HOW TO GATHER INFORMATION

You can gather information for your analysis by:

PRAYER:

Ask God to reveal what you need to know about this area in order to reach it with the Gospel.

STUDY OF WRITTEN MATERIALS:

Study books and written materials about the area. A list of possible sources of information is included in the Appendix of this manual.

OBSERVATION:

Travel personally throughout the area and observe its land and peoples.

INTERVIEW:

Ask questions of people who are leaders and residents of the area. If there are believers already ministering in the area, interview them to discover what they have learned. Sample interviews are included in the Appendix of this course.

Use the following outline to analyze an area. You may not be able to answer all of the questions about an area, but answer those for which you can obtain information.

IDENTIFYING AN AREA

First, be certain the territory you have selected is a distinct area. An area is considered distinct or separate for at least one of the following reasons:

-The people in the area all speak the same language.

-The people groups in the area are all similar culturally. For example, all the people groups in the area are all tribal groups, or they all may be members of a certain religion.

-The area is divided from the rest of the nation by natural geographic elements such as mountains, a river, etc.

-The area is separate politically: It can be a...

-Village
-Township
-County

-State
-Province
-Separate territory owned by a country
-Other:________________

WHAT PEOPLE GROUPS ARE WITHIN THIS AREA?

1. What people groups live in this area? Obtain the name and approximate population of each group in the area.

2. Obtain or draw a map of the area and locate each people group on it. Begin to pray over this area and its different people groups.

3. What is the total population of all the people in the whole area?

WHAT DO THESE GROUPS HAVE IN COMMON?

In general, how are these groups alike:

a. Socially:
b. Culturally:
c. Linguistically:
d. Religiously:
e. Politically:

How do these groups differ:

a. Socially:
b. Culturally:
c. Linguistically:
d. Religiously:
e. Politically:

(Note: You may want to do a "People Group Analysis" on each different group in the area. This will help you understand and design a strategy to reach them.)

WHAT IS THIS AREA LIKE?

GEOGRAPHICAL ANALYSIS:

1. What geographic factors will hinder the spread of the Gospel in this area?

2. What geographic factors might be used to help the spread of the Gospel in this area? (For example, are there natural routes of trade or travel along which churches could be planted?)

3. Does the weather pattern make certain times of the year best for traveling throughout the area to minister?

4. Obtain information on the following factors which will assist you in reaching this area:

    a. Major cities: Mark them on your map. In these areas the population is largest, so you will want to be sure to design a strategy to reach them in your plan.

    b. Established airports: Mark them on your map.

    c. Usable airstrips or water landings: Mark them on your map.

    d. Does the area have telephone or Internet service?

    e. Does the area have telegraph service?

    f. Does the area have rail or bus transportation?

    g. Are there usable automobile roads? Locate them on your map.

    h. Are there navigable rivers or seas? ("Navigable" means that you can use a boat to travel). Locate them on your map.

SOCIAL ANALYSIS:

1. How open are the people to "outsiders" (people who are not from the same area)? Knowing this will prepare you for the level of acceptance you will experience when you first enter this area to share the Gospel.

2. In general, who controls the financial and material resources of the area? You may need their assistance to obtain property for a church and finances for spreading the Gospel.

CULTURAL ANALYSIS:

Study cultural practices of this area that may affect acceptance of the Gospel. Here are some to consider:

1. Beliefs And Values:

    a. What cultural values, beliefs, or practices of the area may help the people understand the Gospel?

b. What cultural values, beliefs, or practices of the people might hinder their understanding of the Gospel?

2. Cultural Practices:

Are there some common cultural practices in this area that are clearly prohibited by Scripture? If so, what are they? You will have to deal with them.

3. Needs:

a. Consider the area as a whole and identify the most important needs:

_____Inadequate food.
_____Poor health and medical services.
_____Poor education.
_____Improper water and/or poor supplies of it.
_____Inadequate financial resources.
_____Inadequate geographic resources.
_____Poor transportation.
_____Poor systems of communication.
_____Inadequate housing.
_____Inability to cope with change.
_____Political oppression.
_____Other:___________________________________

Your presentation of the Gospel and your methods of ministry in this area can relate to these needs.

b. What needs appear to be well met in this area? What needs does the area not have?

4. Education:

What is the general educational level of the majority of the people in this area? What is the literacy level? How many of the people read and write their own language? (It is important to know this because you would not want to design a Gospel outreach based on literature if most people in an area do not read and write.)

You might also research:

a. How many elementary schools are in the area?
b. How many secondary schools are in the area?
c. How many colleges are in the area?

d. How many Bible schools are in the area?

(Educational programs may be one method you can use to reach this area if there is a need for additional schools.)

5. Communication:

a. Which language (or languages) is most suitable for presentation of the Gospel in this area? You want to reach the largest number of people in an area first, so select as first choice the language which most of the people speak.

The main language spoken by people in this area is________.

Other languages needed later to reach the remainder of the people:

____

b. What is the status of Scripture translation in the language that is most suitable for reaching this area?
New Testament:

Old Testament:

c. Can the people in this area be reached through:

_____Newspapers (Do they accept Christian advertisements?)
_____Printed literature (Can the people to be reached read?)
_____Radio (Do the stations permit Christian broadcasting?)
_____Television (Do the stations permit Christian broadcasting?)
_____Tape recordings (Do the people have access to tape recorders?)
_____Movies, video tapes (Is the equipment available in this area?)
_____Telephone (Do the people have telephones or Internet service?)

d. What is the main method of communication throughout the area?

EXTERNAL INFLUENCES:

1. What changes are taking place in this area as a result of the political system or outside forces of modernization?

2. Can any of these changes be used to help spread the Gospel?

3. What organized groups or agencies from outside the area may oppose the spread of the Gospel within this area? Why?

Never let restrictions or opposition prevent you from ministering to an area to which you feel God has called you. If you are not allowed to reach them in certain ways, use other methods.

POLITICAL ANALYSIS:

1. Briefly describe the political system in this area.

2. What political groups in this area might hinder the spread of the Gospel?

3. What political groups in this area might help in the spread of the Gospel?

4. What government regulations might hinder your ministry in this area?

5. What government regulations might assist in your ministry in this area?

6. Who are the key political leaders in this area? You might make their acquaintance and try to gain their favor for your ministry in the area.

SPIRITUAL ANALYSIS:

1. What religious groups are found in this area?

2. How many of the people follow each religion?

3. What is the general attitude in this area towards religious change of any kind? The answer to this question will prepare you for religious opposition you might be facing.

4. What is the total number of professing Christians?

5. Can you identify key Christian leaders in this area?

6. Do you consider this area:

   _____Totally unevangelized.
   _____Partially evangelized.
   _____Adequately evangelized.

7. In general, do you believe this is a time for sowing the Gospel in this area or a time for reaping?

8. What Christian churches or missions have been attempted in this area and failed? What do you believe were the reasons for their failure? You can learn from the failures of others and not repeat the same mistakes.

9. What Christian churches or missions are now active within the area?

10. Which of these are effective and why are they effective? You can learn from the successes of these groups. You will also want to cooperate with them in spreading the Gospel in this area.

11. How can you cooperate with them to reach this area?

12. Among which people groups are the existing ministries working?

13. Are there any people groups in this area who do not yet have a Christian witness among them?

14. If there is a Christian witness present in this area, what language or languages are being used? You may be led to reach people in this area who speak a different language and are not currently being reached by the present Christian witness.

15. What types of Christian ministry are presently being done in this area? Do not duplicate or compete with existing efforts when you formulate your plan. Your goal is to reach the unreached with the Gospel.

16. What is the general attitude of people in this area toward these existing Christian missions?

    _____Hostile
    _____Indifferent
    _____Friendly

17. How many true Gospel churches are in this area?

18. How fast are they growing?

19. What difficulties are they experiencing and why?

20. Which ones are successful and why?

21. Is there a need to establish more churches?

22. Can you determine what principalities and powers of Satan are active or ruling this area? (Ruling principalities and powers of Satan were explained in Chapter Six.)

WHO SHOULD REACH THIS AREA?

Who should reach this area? How many workers are needed to adequately reach this area? Your population an estimates will help you determine this.

What are the necessary qualifications for workers to spread the Gospel in this area? Whoever goes to share the Gospel in this area should...

-Be a believer with evidence of this in their lifestyle, attitudes, and behavior.

-Have good personal prayer and Bible study habits.

-Have spiritual gifts properly suited to the methods to be used to reach the people of the area.

-Be available: Ability is no good without availability. Is the person willing and ready to go?

-Be properly fitted into the body (willing to cooperate with other members for the goal of reaching this area with the Gospel).

-Be called of God to minister to this specific area.

-Be adequately prepared physically, spiritually, and educationally to reach this area.

HOW SHOULD THIS AREA BE REACHED?

How should this area be reached? What methods should be used? To answer this question, use what you learned in your geographic, linguistic, cultural, political, and spiritual analysis of this area. What similarities between the people groups in this area permit common methods to be used? What differences between the people groups call for different methods? Be sure your methods are Scriptural. Do not use any method that conflicts with Biblical principles.

Which evangelistic strategies have been effective in this area? What new methods might you try? After you have determined the methods by prayer and analysis, ask this question: What opposition may be raised against these methods by either the people groups in this area or external forces? How can you deal with this opposition?

FORMULATE A PLAN

The analysis you have completed is of no benefit unless you use what you learned about the area you studied. This area will not be reached by study alone. You must make a plan to reach it with the Gospel. Chapter Eleven, "A Strategy For Planning," will help you formulate a plan to reach the area you have analyzed.

SELF-TEST

1. Write the Key Verse from memory.

____

____

2. Define the term "area" as it is used in this lesson.

____

____

3. What does an area analysis include?

____

____

4. What are the purposes of an area analysis?

____

____

5. As part of your study of this lesson, did you actually complete the analysis of an area?

____

6. Have you made a plan to use what you learned to spread the Gospel within this area?

____

(Answers to tests are provided at the conclusion of the final chapter in this manual.)

FOR FURTHER STUDY

Study Genesis chapter 13 which provides an analysis of two different areas. One was "The Plains Of Jordan," the other was the "Land Of Canaan." Record the results of your analysis:

Plains Of Jordan	**Land Of Canaan**

CHAPTER EIGHT

ANALYZING A NATION

OBJECTIVES:

Upon completion of this chapter you will be able to:

- Write the Key Verse from memory.
- Define the term "nation."
- Explain what an analysis of a nation includes.
- Identify purposes of an analysis of a nation.
- Analyze a nation.
- Use your analysis to spread the Gospel within this nation.

KEY VERSE:

And that repentance and remission of sins should be preached in His name among all nations beginning at Jerusalem. (Luke 24:47)

INTRODUCTION

You have learned how to analyze a people group and a specific area within a nation. In this lesson you will learn how to analyze an entire nation for the purpose of reaching it with the Gospel.

A NATION

A nation is composed of people united by a common political system. It is a territory with established national boundaries.

ANALYSIS OF A NATION

When you analyze a nation you study it in detail to determine political, geographical, linguistic, cultural, spiritual factors and external forces that might affect the spread of the Gospel throughout the nation.

THE PURPOSE OF ANALYZING A NATION

The purpose of your study is to expand from an area in which you are ministering to reach an entire nation with the Gospel. It is research for the purpose of evangelism, not just for the sake of research. Be careful that you do not spend all of your time studying and never get around to accomplishing the purpose which originally motivated your study! The things which unite a nation and make it distinct from other nations can provide us with a general strategy to reach the entire nation with the Gospel.

You analyze a nation to answer these questions:

WHAT PEOPLE GROUPS ARE WITHIN THIS NATION?

Who are they, how many, and where are they located? Which ones are reached with the Gospel and which ones are not?

WHAT DO THESE GROUPS HAVE IN COMMON?

What unites these people groups as part of this nation? How are they alike socially, culturally, linguistically, religiously, and politically?

WHAT IS THIS NATION LIKE?

What do we need to know about this nation in order to reach the people who live there with the Gospel?

WHO SHOULD REACH THIS NATION?

What are the necessary qualifications for workers to spread the Gospel in this nation?

HOW SHOULD THIS NATION BE REACHED?

What general methods can be used in the whole nation? Which will be most effective? The answers to these questions will help you determine your role in reaching a nation and assist you in formulating a plan to reach it with the Gospel.

HOW TO GATHER INFORMATION

You can gather information for your analysis by:

PRAYER:

Ask God to reveal what you need to know about this nation in order to reach it with the Gospel.

STUDY OF WRITTEN MATERIALS:

Study books and written materials about the nation. A list of possible sources of information is included in the Appendix of this manual.

OBSERVATION:

Travel personally throughout the nation and observe its land and peoples.

INTERVIEW:

Ask questions of people who are leaders and citizens of the nation. If there are believers already ministering in the nation, interview them to discover what they have learned. Sample interviews are included in the Appendix of this course.

Use the following outline to analyze a nation. You may not be able to answer all of the questions about a nation, but answer those for which you can obtain information.

WHAT PEOPLE GROUPS ARE WITHIN THIS NATION?

1. What people groups live in this nation? Obtain the name and approximate population of each group in the nation.

2. Obtain or draw a map of the nation and locate each people group on it. Does the nation divide into several distinct areas based on the location of various people groups? Begin to pray over the nation and its different people groups.

3. What is the total population of all the people in the whole nation?

WHAT DO THESE GROUPS HAVE IN COMMON?

How are the various people groups in this nation alike:

a. Socially:
b. Culturally:
c. Linguistically:
d. Religiously:
e. Politically:

How do the people groups in this nation differ:

a. Socially:
b. Culturally:

c. Linguistically:
d. Religiously:
e. Politically:

(Note: You may want to do a People Group Analysis on each different group in the nation. This will help you design a strategy to reach each individual group.)

WHAT IS THIS NATION LIKE?

GEOGRAPHICAL ANALYSIS:

1. What geographic factors will hinder the spread of the Gospel in this nation?

2. What geographic factors might be used to help the spread of the Gospel in this nation? For example, are there natural routes of trade or travel along which churches could be planted?

3. Does the weather pattern make certain times of the year best for ministering in certain areas of the nation?

4. Divide the nation into areas based on natural geographic factors. How will geography affect your strategy in each area?

5. Obtain information on the following factors which will assist you in reaching this nation:

   a. Major cities: Mark them on your map. The population is largest in these cities, so you will want to be sure to design a strategy to reach them in your plan.

   b. Established airports: Mark them on your map.

   c. Usable airstrips or water landings: Mark them on your map.

   d. What areas of the nation have telephone or Internet service?

   e. What areas have telegraph service?

   f. What areas have rail or bus transportation?

   g. To what areas of the nation will you have to travel by foot or horseback?

   h. Where are the usable automobile roads? Locate them on your map.

   I. Are there navigable rivers or seas? ("Navigable" means that you can use a boat to travel). Locate them on your map.

SOCIAL ANALYSIS:

1. How open are the people of this nation to "outsiders"? Knowing this will prepare you for the level of acceptance you will experience when you first enter this nation to share the Gospel.

2. In general, who controls the financial and material resources of this nation? You may need their assistance to obtain property for a church and finances for spreading the Gospel.

CULTURAL ANALYSIS:

Study cultural practices of this area that may affect acceptance of the Gospel. Here are some to consider:

1. Beliefs And Values:

   a. What cultural values, beliefs, or practices of the nation may help the people understand the Gospel?

   b. What cultural values, beliefs, or practices may hinder their understanding of the Gospel?

2. Cultural Practices:

   Are there some common cultural practices in this nation that are clearly prohibited by Scripture? If so, what are they? You will have to deal with them.

3. Needs:

   a. Consider the nation as a whole and identify the most important needs:

      _____Inadequate food.
      _____Poor health and medical services.
      _____Poor education.
      _____Improper water and/or poor supplies of it.
      _____Inadequate financial resources.
      _____Inadequate geographic resources.
      _____Poor transportation.
      _____Poor systems of communication.
      _____Inadequate housing.
      _____Inability to cope with change.
      _____Political oppression.
      _____Other:____________________________________

Your presentation of the Gospel and perhaps your methods of ministry in this nation can relate to these needs.

b. What needs appear to be well met in this nation? What needs does the nation not have?

4. Education:

What is the general educational level of the majority of the people in this nation? What is the literacy level? How many of the people read and write their own language? It is important to know this because you would not want to design a Gospel outreach based on literature if most people in a nation do not read and write.

You might also research:

a. Is there adequate number of elementary schools throughout the nation?
b. Is there an adequate number of secondary schools throughout the nation?
c. How many colleges are in the nation?
d. How many Bible schools are in the nation?

Educational programs may be one method you can use to reach this nation if there is a need for additional schools.

5. Communication:

a. Which language (or languages) is most suitable for presentation of the Gospel in this nation? You want to reach the largest number of people in an area first, so select as first choice the language which most of the people speak.

The main language spoken in this nation:________________

Other languages needed later to reach the remainder of the people:

____

b. What is the status of Scripture translation in the language that is most suitable for reaching this nation?

New Testament:

Old Testament:

c. Do a similar analysis of the status of Scripture translation for the other languages in the nation.

Language:___________________

New Testament:

Old Testament:

d. On your map, note the location of the various linguistic divisions in the country. In what areas do they speak the various languages?

e. Can the majority of the people in this nation be reached through:

_____Newspapers (Do they accept Christian advertisements?)
_____Printed literature (Can the majority of the people read?)
_____Radio (Do the stations permit Christian broadcasting?)
_____Television (Do the stations permit Christian broadcasting?)
_____Tape recordings (Do the people have access to tape recorders?)
_____Movies, video tapes (Is the equipment available in this nation?)
_____Telephone (Do the majority of the people have telephones or Internet?)

f. What is the main method of communication common throughout the nation?

EXTERNAL INFLUENCES:

1. What changes are taking place in this nation as a result of the political system or outside forces of modernization?

2. Can any of these changes be used to help spread the Gospel?

3. What organized groups or agencies from outside the nation may oppose the spread of the Gospel within this nation? Why?

   Never let restrictions or opposition prevent you from ministering in a nation to which you feel God has called you. If you are not allowed to reach them in certain ways, use other methods.

POLITICAL ANALYSIS:

1. Briefly describe the political system in this nation:

2. What political groups in this nation might hinder the spread of the Gospel?

3. What political groups in this nation might help in the spread of the Gospel?

4. What government regulations might hinder your ministry in this nation?

5. What government regulations might assist in your ministry in this nation?

6. Who are the key political leaders in this nation? You might want to make their acquaintance and gain their favor for your nationwide ministry.

SPIRITUAL ANALYSIS:

1. What religious groups are found in this nation?

2. How many of the people follow each religion?

3. What is the general attitude in this nation towards religious change of any kind? The answer to this question will prepare you for religious opposition you might be facing.

4. What is the total number of professing Christians?

5. Can you identify key Christian leaders?

6. Do you consider this nation:

   _____Totally unevangelized.
   _____Partially evangelized.
   _____Adequately evangelized.

7. In general, do you believe this is a time for sowing the Gospel in this nation or a time for reaping?

8. What Christian churches or missions have been attempted in this nation and failed? What do you believe were the reasons for their failure? You can learn from the failures of others and not repeat the same mistakes.

9. What significant Christian churches or missions are now active within the nation?

10. Which of these are effective and why are they effective? You can learn from the successes of these groups. You will also want to cooperate with them in spreading the Gospel in this nation.

11. How can you cooperate with existing ministries to reach this nation?

12. Among which people groups are the existing ministries working?

13. Are there any people groups in this nation who do not yet have a Christian witness among them?

14. Are there any language groups that do not yet have a Christian witness?

15. What types of Christian ministry are presently being done in this nation? Do not duplicate or compete with existing efforts when you formulate your plan. Your goal is to reach the unreached with the Gospel.

16. What is the general attitude of people in this nation towards these existing Christian missions?

    _____Hostile
    _____Indifferent
    _____Friendly

17. How many true Gospel churches are in this nation?

18. How fast are they growing?

19. What difficulties are they experiencing and why?

20. Which ones are successful and why?

21. Is there a need to establish more churches?

22. Can you determine what principalities and powers of Satan are active or ruling this nation? (Ruling principalities and powers of Satan were explained in Chapter Six.)

WHO SHOULD REACH THIS NATION?

Who should reach this nation? How many workers are needed to adequately reach this nation? Your population estimates will help you determine this. What are the necessary qualifications for workers to spread the Gospel in this nation? Whoever goes to share the Gospel in this nation should...

-Be a believer with evidence of this in their lifestyle, attitudes, and behavior.

-Have good personal prayer and Bible study habits.

-Have spiritual gifts properly suited to the methods to be used to reach the people of the nation.

-Be available: Ability is no good without availability. Is the person willing and ready to go?

-Be properly fitted into the body (willing to cooperate with other members for the goal of reaching this nation with the Gospel).

-Be called of God to minister to this specific nation.

-Be adequately prepared physically, spiritually, and educationally to reach this nation.

HOW SHOULD THIS NATION BE REACHED?

How should this nation be reached? What methods should be used? To answer this question, use what you learned in your geographic, linguistic, cultural, political, and spiritual analysis of this nation.

What similarities between the people groups in this nation permit common methods to be used? What differences between the people groups call for different methods?

How could the nation be divided into smaller areas and then each area reached with the Gospel. (Perhaps you will want to do an area analysis on the major areas within the nation.)

Be sure your methods are Scriptural. Do not use any method that conflicts with Biblical principles. Refer to the list on pages 75-76 of various methods that have been used as part of evangelism strategies in different areas of the world. Which of these might be effective in this nation? What new methods might you try?

After you have determined the methods by prayer and analysis, ask this question: What opposition may be raised against these methods by either internal or external forces? How can you deal with this opposition?

FORMULATE A PLAN

The analysis you have completed is of no benefit unless you use what you learned about the nation you studied. This nation will not be reached by study alone. You must make a plan to reach it with the Gospel. Chapter Eleven, "A Strategy For Planning," will help you formulate a plan to reach the nation you have analyzed.

SELF-TEST

1. Write the Key Verse from memory.

____

____

2. Define the term "nation."

____

3. What does an analysis of a nation include?

____

4. What are the purposes of an analysis of a nation?

____

____

5. As part of your study of this lesson, did you actually complete the analysis of a nation?

____

6. Have you made a plan to use what you learned to spread the Gospel throughout this nation?

____

(Answers to tests are provided at the conclusion of the final chapter in this manual.)

FOR FURTHER STUDY

Study the historical and spiritual analysis of the nation of Israel in Acts chapter 7.

CHAPTER NINE

ANALYZING A REGION

OBJECTIVES:

- Upon completion of this chapter you will be able to:
- Write the Key Verse from memory.
- Define the term "region."
- Explain what a regional analysis includes.
- Identify purposes of a regional analysis.
- Analyze a region.
- Use your analysis to spread the Gospel within this region.

KEY VERSE:

> **And He said unto them, Go ye into all the world, and preach the Gospel to every creature. (Mark 16:15)**

INTRODUCTION

You have learned how to analyze a people group, an area, and a nation. In this lesson you will learn how to analyze an entire region of the world for the purpose of reaching it with the Gospel.

A REGION

For purposes of this study, a region is a part of the world made up of several nations located near each other geographically which are similar linguistically, culturally, religiously, or politically.

The Central American region is a good example. The nations in this region are Guatemala, Nicaragua, Costa Rica, Honduras, Panama, Belize, and San Salvador. These nations are in a similar geographic location, share similar cultures and religions, and Spanish is the main language spoken.

Just because nations are in a similar geographic location does not make them "regions" for purposes of our study. For example, Communist China is near the Philippine islands, but these two regions are very different. Their language, culture, religions, and political systems are very different. Greatly different methods of evangelism are necessary in these two regions.

In our regional analysis we are looking for regions that are similar in culture, politics, religion, and language like the Central American Region. Our purpose is to design similar plans to reach them with the Gospel.

ANALYSIS OF A REGION

When you analyze a region you study it in detail to determine political, geographical, linguistic, cultural, spiritual factors and external forces that might affect the spread of the Gospel throughout the region.

THE PURPOSE OF ANALYZING A REGION

The purpose of your study is to expand from the nations in which you are ministering to reach an entire region with the Gospel.

A regional analysis is research for the purpose of evangelism, not just for the sake of research. Be careful that you do not spend all of your time studying and never get around to accomplishing the purpose which originally motivated your study!

The things which unite a region and make it distinct from other regions can help us plan a general strategy to reach the entire region with the Gospel. The world is made up of several regions similar to the Central American region which was used as an example. If we reach enough regions, eventually we will reached the entire world.

You analyze a region to answer these questions:

WHAT NATIONS AND PEOPLE GROUPS ARE WITHIN THIS REGION?

What nations are within this region? How many nations, what is their population, and what makes them a region?

WHAT DO THESE NATIONS AND PEOPLE GROUPS HAVE IN COMMON?

How are they alike socially, culturally, linguistically, religiously, and politically?

WHAT IS THIS REGION LIKE?

What do we need to know about this region in order to reach the people who live there with the Gospel?

WHO SHOULD REACH THIS REGION?

What are the necessary qualifications for workers to spread the Gospel in this region? How many workers are needed?

HOW SHOULD THIS REGION BE REACHED?

What similar methods can be used in the whole region? Which will be most effective?

The answers to these questions will help you determine your role in reaching a region and assist you in making a plan to reach it with the Gospel.

HOW TO GATHER INFORMATION

You can gather information for your analysis by:

PRAYER:

Ask God to reveal what you need to know about this region in order to reach it with the Gospel.

STUDY OF WRITTEN MATERIALS:

Study any existing books and written materials about the region. A list of possible sources of information is included in the Appendix of this manual.

OBSERVATION:

Travel personally throughout the region. Visit each nation and observe its land and peoples.

INTERVIEW:

Ask questions of people who are leaders in the nations of this region. If there are believers who already have regional ministries in the region you are studying, interview them to discover what they have learned. Sample interviews are included in the Appendix of this course.

Use the following outline to analyze a region. You may not be able to answer all of the questions about a region, but answer those for which you can obtain information.

IDENTIFYING A REGION

First make sure you properly identify a region. A region consists of nations...

-Which are located near each other geographically.
-Where the majority of the people speak the same language.
-Which have similar political systems.
-Where the majority of the people follow a common religion.
-Where the nations have similar political systems.
-Where the majority of the people have a similar culture.

The nations may not be alike in all of these, but a region should definitely consist of nations located near each other geographically. It should also be alike in at least two or more of the other options on the list in order to plan a common strategy to reach it with the Gospel.

WHAT NATIONS AND PEOPLE GROUPS ARE WITHIN THIS REGION?

1. What nations and major people groups live in this region? Identify the name and approximate population of each nation and/or major people group. (You may need to analyze each nation to answer this question. Use Lesson Eight, "Analyzing A Nation.")

2. Locate the major people groups on your regional map. Does the region divide into several distinct areas based on the location of various people groups? You may want to design a plan to reach this region by major people groups. Begin to pray over this region and its different people groups. (You may want to analyze these groups individually. Use Lesson Six, "Analyzing A People Group.")

3. What is the total population of all the people in the various nations? In the entire region?

WHAT DO THESE NATIONS AND GROUPS HAVE IN COMMON?

How are the various nations and people groups in this region alike:

a. Socially:
b. Culturally:
c. Linguistically:
d. Religiously:
e. Politically:

How do the nations and people groups in this region differ:

a. Socially:
b. Culturally:
c. Linguistically:
d. Religiously:
e. Politically:

WHAT IS THIS REGION LIKE?

GEOGRAPHICAL ANALYSIS:

1. What geographic factors will hinder the spread of the Gospel in this region?

2. What geographic factors might be used to help the spread of the Gospel in this region?

3. Does the weather pattern make certain times of the year best for ministering in certain nations of the region?

4. Obtain a map and note the following factors which will assist you in reaching this region:

   a. Nations: Identify the various nations in the region. In each nation, identify the following:

   b. Capitols and major cities: Mark them on your map. The population is largest in these cities, so you will want to be sure to design a strategy to reach them in your plan.

   c. Established airports: Mark them on your map.

   d. Usable airstrips or water landings: Mark them on your map.

   e. What nations in this region have telephone or Internet service?

   f. What nations have telegraph service?

   g. What nations have good rail or bus transportation?

   h. Where are the usable automobile roads? Locate them on your map.

   I. Are there navigable rivers or seas? ("Navigable" means that you can use a boat to travel). Locate them on your map.

SOCIAL ANALYSIS:

1. In general, which nations in this region are the most open to "outsiders"? Knowing this will prepare you for the level of acceptance you will experience when you first enter this region to share the Gospel. You may also want to target the most receptive nations of the region first in your plan to reach them with the Gospel.

2. In general, who controls the financial and material resources of this region?

CULTURAL ANALYSIS:

1. Beliefs And Values:

   a. What cultural values, beliefs, or practices of this region may help the people understand the Gospel?

b. What cultural values, beliefs, or practices may hinder their understanding of the Gospel?

2. Cultural Practices:

Are there some common cultural practices in this region that are clearly prohibited by Scripture? If so, what are they? You will have to deal with them.

3. Needs:

a. Are there any needs common to all the nations in this region?

_____Inadequate food.
_____Poor health and medical services.
_____Poor education.
_____Improper water and/or poor supplies of it.
_____Inadequate financial resources.
_____Inadequate geographic resources.
_____Poor transportation.
_____Poor systems of communication.
_____Inadequate housing.
_____Inability to cope with change.
_____Political oppression.
_____Other:_______________________________

If there are certain needs common to all the nations in the region, you may want to use that need as a key to reach the region with the Gospel.

b. What needs appear to be well met in this region? What needs does the region not have?

4. Education:

What is the general educational level of the majority of the people in this region? What is the general literacy level? How many of the people read and write the language you will be using in this region?

It is important to know this because you would not want to design a Gospel outreach based on literature if most people in a region do not read and write.

Is there a general need in education common throughout the region. For example, is there a general lack of elementary, secondary, college, or Bible schools? Educational programs may be one method you can use to reach the region if there is a common need for additional schools.

5. Communication:

   a. Which language (or languages) is most suitable for presentation of the Gospel in this region? You want to reach the largest number of people in a region first, so select as first choice the language which most of the people speak.

The main language spoken by people in this region is____________.

Other languages needed later to reach the remainder of the people: __________________.

   b. What is the status of Scripture translation in the language that is most suitable for reaching this region?

   New Testament:

   Old Testament:

   c. Can the majority of the people in this region be reached through:

   _____Newspapers (Do they accept Christian advertisements?)
   _____Printed literature (Can the majority of the people read?)
   _____Radio (Do the stations permit Christian broadcasting?)
   _____Television (Do the stations permit Christian broadcasting?)
   _____Tape recordings (Do the people have access to tape recorders?)
   _____Movies, video tapes (Is the equipment available in this region?)
   _____Telephone (Do the majority of the people in this region have telephones or Internet service?)

   d. What is the main method of communication common throughout the region?

EXTERNAL INFLUENCES:

1. What changes are taking place in this region as a result of outside political forces or modernization?

2. Can any of these changes be used to help spread the Gospel?

3. What nations from outside of this region may oppose the spread of the Gospel within the region? Why?

Never let restrictions or opposition prevent you from ministering in a region to which you feel God has called you. If you are not allowed to reach them in certain ways, use other methods.

POLITICAL ANALYSIS:

1. Analyze the political system in this region. How is it similar in each nation of the region? How is it different?

2. Which nations have political systems that might hinder the spread of the Gospel?

3. Which nations have political systems that might help in the spread of the Gospel?

4. What regulations are common to the region which might hinder your ministry?

5. What common government regulations might assist in your ministry in this region?

6. Who are the key political leaders in this region? You make their acquaintance and/or try to gain their favor for your regional ministry.

SPIRITUAL ANALYSIS:

1. What religious groups are common to the nations in this region?

2. How many of the people follow these religions?

3. What is the general attitude of people in the region towards religious change of any kind? The answer to this question will prepare you for religious opposition you might be facing.

4. What is the total number of professing Christians in the region?

5. Can you identify key Christian leaders in this region?

6. Do you consider this region:

   _____Totally unevangelized.
   _____Partially evangelized.
   _____Adequately evangelized.

7. In general, do you believe this is a time for sowing the Gospel in this region or a time for reaping?

8. What Christian churches or missions have been attempted in this region and failed?

What do you believe were the reasons for their failure? You can learn from the failures of others and not repeat the same mistakes.

9. What significant Christian churches or missions are now active within this region?

10. Which of these are effective and why are they effective? You can learn from the successes of these groups. You will also want to cooperate with them in spreading the Gospel in this region.

11. How can you cooperate with existing ministries who are attempting to reach this region?

12. In which nations and among which people groups are the existing ministries working?

13. Are there any nations or people groups which do not yet have a Christian witness among them?

14. Are there any language groups of the region that do not yet have a Christian witness?

15. What types of Christian ministry are presently being done in this region? Do not duplicate or compete with existing efforts when you formulate your plan. Your goal is to reach the unreached with the Gospel.

16. What is the general attitude of people in this region towards these existing Christian missions?

_____Hostile
_____Indifferent
_____Friendly

17. How many true Gospel churches are in this region?

18. How fast are they growing?

19. What difficulties are they experiencing and why?

20. Which ones are successful and why?

21. Is there a need to establish more churches?

22. Can you determine what principalities and powers of Satan are active or ruling this region? (Ruling principalities and powers of Satan were explained in Chapter Six).

WHO SHOULD REACH THIS REGION?

Who should reach this region? How many workers are needed to adequately reach this nation? Your population estimates will help you determine this.

What are the necessary qualifications for workers to spread the Gospel in this region? Whoever goes to share the Gospel in this region should...

-Be a believer with evidence of this in their lifestyle, attitudes, and behavior.

-Have good personal prayer and Bible study habits.

-Have spiritual gifts properly suited to the methods to be used to reach the nations of the region.

-Be available: Ability is no good without availability. Is the person willing and ready to go?

-Be properly fitted into the body (willing to cooperate with other members for the goal of reaching this region with the Gospel).

-Be called of God to minister to this specific region.

-Be adequately prepared physically, spiritually, and educationally to reach this region.

HOW SHOULD THIS REGION BE REACHED?

How should this region be reached? What similar methods can be used in the whole region? Which will be most effective? To answer this question, use what you learned in your geographic, linguistic, cultural, political, and spiritual analysis of this region.

What similarities between the nations of this region permit common methods to be used? What differences between the nations of this region require different methods to be used?

Be sure your methods are Scriptural. Do not use any method that conflicts with Biblical principles. Refer to the list of various methods that have been used as part of evangelistic strategies in different areas of the world provided in Chapter Eight. Which of these might be effective in this region? What new methods might you try?

After you have determined the methods by prayer and analysis, ask this question: What opposition may be raised against these methods by either internal or external forces? How can you deal with this opposition?

FORMULATE A PLAN

The analysis you completed will be of no benefit unless you use what you learned about the region you studied. This region will not be reached by study alone. You must make a plan to reach it with the Gospel.

Study Chapter Eleven, "A Strategy For Planning," to help you make a plan to reach the region you have analyzed.

SELF-TEST

1. Write the Key Verse from memory.

____

____

2. Define the term "region."

____

____

3. What does a regional analysis include?

____

____

4. What are the purposes of a regional analysis?

____

____

5. As part of your study of this lesson, did you actually complete the analysis of a region?

____

6. Have you made a plan to use what you learned to spread the Gospel throughout this region?

____

(Answers to tests are provided at the conclusion of the final chapter in this manual.)

FOR FURTHER STUDY

The following REAP plan of organization will achieve the purpose of extending the Gospel to the nations of the world.

R...Regions:	Divide the world into major geographical regions. Appoint a REGIONAL COORDINATOR to serve each region.
E...Every Nation:	Every nation within each region must be reached. A NATIONAL DIRECTOR should be appointed over each nation to serve under the Regional Coordinator.
A...Areas Within Nations:	Because of the cultural, linguistic, and geographical diversity within nations, each nation should be divided into areas on the basis of these factors. An AREA REPRESENTATIVE will serve under the National Director.
P...People Groups:	The modern missionary movement is viewing the challenge of the unreached in terms of people groups. People groups can be represented by a GROUP SUPERVISOR.

The REAP plan of organization follows the organizational plan of extension given by the Lord Jesus Christ:

> **...and ye shall be witnesses unto me both in Jerusalem, and in all Judaea, and in Samaria, and unto the uttermost part of the earth. (Acts 1:8)**

The plan of Jesus included regions, each nation, areas, and people groups:

Place	**Representative Of**
Uttermost parts of the earth:	Regions
Samaria:	Every nation
Judea:	Area around Jerusalem
Jerusalem:	People groups

CHAPTER TEN

ANALYZING AN EXISTING CHRISTIAN ORGANIZATION

OBJECTIVES:

Upon completion of this chapter you will be able to:

- Write the Key Verse from memory.
- Explain what an analysis of a Christian organization includes.
- Identify purposes for analyzing a Christian organization.
- Analyze an existing Christian organization.
- Use the results of your analysis to help the organization fulfill its unique purpose in God's plan.

KEY VERSE:

> **He that hath an ear, let him hear what the Spirit saith unto the churches. (Revelation 2:29)**

INTRODUCTION

All true believers who are born again are part of the one true Church which is called the Body of Christ:

> **For by one Spirit are we all baptized into one body... (I Corinthians 12:13)**

(John 3 describes this experience of being "born again" by the Spirit.)

This "Body of Christ" or Church has been organized in different ways throughout the world. There are many local churches, mission agencies, denominations, fellowships, and Christian service organizations. These various organizations have been developed to better fulfill the purposes of the Church at a local level. In this lesson you will learn how to analyze such Christian organizations.

THE PURPOSE OF ANALYZING AN ORGANIZATION

It is God's plan that the Church be the spiritual force to spread the Gospel to the nations of the world:

> **...To make all men see what is the fellowship of the mystery, (the Gospel) which from the beginning of the world hath been hid in God, who created all things by Jesus Christ.**

To the intent that now unto the principalities and powers in heavenly places might be known by the Church the manifold wisdom of God.

According to the eternal purpose which He purposed in Christ Jesus our Lord. (Ephesians 3:9-11)

The Bible indicates that the spiritual labor force to accomplish this mission is limited:

Then saith He unto His disciples, The harvest truly is plenteous, but the laborers are few. (Matthew 9:37)

If laborers are few, then they must be organized in the best way possible in order for the Church to successfully fulfill its mission to the world.

You analyze existing Christian organizations for these purposes:

1. To identify the purpose of the organization, its unique part in fulfilling the commission of the Church to spread the Gospel.

2. To determine the present effectiveness of the organization. How well is it currently fulfilling its purpose? How is it succeeding? How is it failing?

3. To formulate plans for the organization to more effectively fulfill its God-given purpose.

If you are a member of the organization which you are analyzing, you will also determine your part in helping the organization achieve its purpose.

WHAT IT INCLUDES

An analysis of an existing organization includes study of its purpose, plans, strengths, weaknesses, environment, resources, and programs.

HOW TO GATHER INFORMATION

You can gather information for your analysis by:

PRAYER:

Ask God to reveal what you need to know about the organization.

STUDY OF WRITTEN MATERIALS:

Study any materials which the organization has in writing. For example, does it have a Doctrinal Statement, Statement of Purpose, written plans, membership requirements, regulations, construction, by-laws, or history?

OBSERVATION:

Personally observe the organization. Watch its members and leaders as they engage in the work of the ministry to which they feel God has called them.

INTERVIEW:

Ask questions of the leaders and members of the organization.

ORGANIZATIONAL ANALYSIS

Here are some things to study when you analyze an organization:

PURPOSE:

1. Does the organization have a written Statement of Purpose? This is a written statement which explains the purpose of the organization. It is not a statement of what the organization believes. It is a statement of why they exist. What is their unique purpose for existing?

2. Do members of the organization understand the purpose? (To find out, ask several members of the organization: "What is the purpose of this organization, church, etc.?")

3. Do members understand their part in fulfilling the purpose of the organization?

4. Is the organization fulfilling its purpose? Is it actually accomplishing what it believes the Lord would have it to do?

5. How does the purpose of the organization relate to God's purpose? (You can read about God's purpose in Ephesians 1:9-11 and 3:9-11).

PLANS:

1. Does the organization have written plans?

2. Do these plans relate to the purpose of the organization? Does each plan contribute towards achieving the purpose of the organization?

3. Do the plans reflect what is to be done, by whom, and by when?

4. Are the plans written so it is possible to tell if they are achieved?

5. Does the organization hold planning meetings? Do they set aside time to pray and seek God for direction?

6. Who makes the plans for the organization?

STRENGTHS:

Analyze the strengths of the organization. What is it doing well? What is good about its present organizational structure?

WEAKNESSES:

Analyze the weaknesses of the organization. Where is it failing? Where does it need improvement?

ENVIRONMENT:

1. In what environment is the organization ministering? (Does it minister to a specific people group, area, nation or region of the world?)

2. Has the organization studied its environment? Has it analyzed the people, area, nation or region to which it is ministering?

3. What things about the environment may change in the next few years?

4. What things will remain the same?

RESOURCES:

Personnel, finances, and facilities are the three main resources of any organization.

1. Personnel: "Personnel" are the people of an organization. Ask these questions:

    a. Are the people of this organization actively using their spiritual gifts to achieve the work of the ministry?

    b. Does each person understand their part in God's plan?

    c. If they are holding an office in the organization do they understand their duties and responsibilities?

d. Is there good communication between the leaders and the people?

e. Is there a need for more workers? If so, in what positions?

f. Is there a good relationship between the people themselves and between the people and the leaders? If not, what are the existing problems? Is there a need for Scriptural discipline or personal counseling to correct these problems?

g. Do leaders in the organization conform to the Biblical pattern of "servant leaders"? (The Harvestime International Institute course, *"Biblical Management Principles,"* discusses servant leadership in detail.)

h. Have the people in the organization received adequate training for the ministry they are attempting to do? How could they be better prepared?

I. How are the people organized? Is there a plan for organizing them to accomplish God's purposes? If laborers are few, then they should be organized in the best possible way to reap the harvest.

j. Some organizations have membership requirements, standards to be met by people who want to join their ministry. If the organization has membership (i.e., church membership) are the requirements for membership Biblical? Do present members live up to these standards?

2. Finances:

It takes money to do God's work. We are responsible for the stewardship of money as well as spiritual gifts. Ask these questions:

a. Is there a written budget? A budget is a plan for how you will spend the money you receive. Such planning is Biblical (See Luke 14:28-30)

b. Does the organization follow this budget?

c. How well is the organization using the funds which God entrusts to it? Are there areas of waste and surplus spending that could be corrected? How could the money be better used to achieve the purposes of the Kingdom of God?

d. Is there a monthly or annual financial report? (A budget is a plan for how the money will be spent. It is made before spending the funds. A financial report shows how the money actually was spent.)

e. If the organization is located in a city, does it have a bank account?

f. Is someone assigned and trained to handle the record of finances, prepare the budget and do financial reports?

g. Does the organization meet all government requirements concerning finances? (For example, annual reports to government, taxes, etc.?)

h. Is the organization violating any Biblical principles in the area of finances?

I. What percentage of the budget goes to evangelism (the actual spreading of the Gospel)?

j. What are the sources of income to this organization? Are there new sources that can be developed?

k. What are the debts that need to be eliminated?

3. Facilities:

Facilities of an organization include buildings and equipment which are used for the work of the ministry. Ask these questions:

a. What are the present building facilities used by or belonging to this organization? Are they adequate for the programs of the organization? How can the present facilities be improved?

b. What new facilities are needed in the future for the organization to adequately fulfill its ministry?

c. If there are government standards for buildings do the present facilities meet these standards? What things need to be corrected? (For example, fire regulations, entrances and exits, emergency equipment, etc.)

d. Make a list of the equipment which belongs to the organization. This list is called an "inventory." Does some of the equipment need repair or replacement?

e. What new equipment is needed to better fulfill their part in God's plan?

f. Where applicable, does the organization have insurance to cover replacement costs of buildings and equipment?

g. Where applicable, what are the costs of heating, cooling, and providing electricity for the facility?

h. Where applicable, is there adequate space for parking cars?

PROGRAMS:

Programs are projects in which an organization engages to accomplish its purposes.

1. Which current programs of the ministry contribute to the unique purpose of the organization? Which do not?

2. Which programs contribute to God's purpose of spreading the Gospel to all nations? Which do not?

3. Which current programs need to be reorganized to be more effective? What needs to be done?

4. Which current programs need to be eliminated?

5. What new programs should be considered which might help the organization better achieve its purpose?

6. Are there any programs in which the organization is engaged which are contrary to Scriptural principles?

7. Are there programs which contribute to the spiritual growth of members of the organization?

MISCELLANEOUS ITEMS:

1. Does the organization have a Doctrinal Statement? This is a written statement of what the organization believes with Biblical references to support each statement of belief.

2. What other problems not previously identified by this analysis need correction?

3. How well does this organization cooperate with other Christian organizations? Are there problems? If so, what are they?

4. If you are a member of the organization you are studying, what is your position in the organization? Are you using your spiritual gifts? What are you doing to help the organization achieve its God-given purpose?

APPLY THE RESULTS OF YOUR ANALYSIS

You must apply the results of your analysis in order for it to be effective. Here is how to do this:

PURPOSE:

1. If there is no written statement as to why the organization exits, how will you know whether or not it achieves its purpose? Work with leaders of the organization to write a Statement of Purpose. Make sure the Statement of Purpose reflects God's purpose of reaching all nations with the Gospel, bringing all things together in Christ.

   To help you write a Statement of Purpose, first review Chapter Eleven of this course and then obtain the Harvestime International Institute course entitled *"Management By Objectives."*

2. When you have a written Statement of Purpose, be sure every member of the church or organization understands the purpose. Explain the unique mission of the organization and the role of each member in accomplishing the purpose.

3. If the organization already has a Statement of Purpose, plan how it can better fulfill the purpose to which God has called it.

STRENGTHS:

Analyze the list of strengths which you identified. How can these best be used to fulfill God's purposes for this organization? How can the strengths be used to reach the world with the Gospel? Strengths that are not used weaken and decay.

WEAKNESSES:

Analyze the list of weaknesses which you identified. How can God work in these weaknesses to demonstrate His power? What Biblical principles might be applied to these weaknesses?

ENVIRONMENT:

If the organization has not analyzed its environment, do an analysis of the people, area, nation, or region to which it is ministering. You must know who you are ministering to if you are to reach them.

Use your analysis of things that may change and what will remain the same in the environment during the next five years to make plans. For example, if a big housing project is scheduled to be built near a certain church, this would be a change in the environment in which they minister. Plans should be made to deal with such changes for the purpose of spreading the Gospel.

RESOURCES:

The Harvestime International Institute course, *"Management By Objectives,"* will help you organize and plan in the following areas:

1. Personnel:

   a. Make a plan to get the people of this organization actively using their spiritual gifts for the work of the ministry.

   b. If the people do not understand their part in the total plan or the duties and responsibilities, this must be corrected. Make a plan of organization.

   c. If there is poor communication between the leaders and the people, make plans for improving communication.

   d. If there is a need for more workers in the organization, begin to recruit and train workers for these ministries.

   e. If there is a poor relationship between the people themselves and/or between the people and the leaders, deal with the problems which you identified. How can you correct these problems? Do principles of Scriptural discipline given in Matthew 18 need to be applied? How can personal counseling be used to help correct these problems?

   f. If the leaders in the organization do not conform to the Biblical pattern of "servant leaders," the Harvestime International Institute course, *"Biblical Management Principles,"* will help in this area.

   g. If people in the organization have not received adequate training for the ministry they are attempting, start a training program. Ask yourself, "Who needs to be better trained? In what do they need training?" Start a training program to meet these needs.

   h. If there are membership requirements for this organization and they are not Biblical, rewrite them.

2. Finances:

   a. If there is no written budget, prepare one.

   b. If the organization is not following its budget as written, then rewrite the budget to more adequately reflect how funds are actually being spent.

c. Eliminate areas of waste or excess spending which you identified.

d. Do a monthly and an annual financial report to summarize how funds are actually spent.

e. If the organization is in a city, start a banking account and ask the bank to help you set up proper systems of accounting for the funds of the organization.

f. Assign and train someone to keep the financial records in an honest and orderly way.

g. If the organization does not meet government requirements for finances, correct this immediately. Submit all required reports, taxes, etc. to the government.

h. If the organization is violating any Biblical principles regarding finances, correct these immediately.

I. Identify ways to increase the percentage of the budget that goes to evangelism (the actual spreading of the Gospel).

j. Identify new sources of income that can be developed for this organization.

k. Make a plan to eliminate debts of the organization.

3. Facilities:

a. Make a list of things in the present facilities which need improvement. For example, what needs repair, paint, new signs, general cleaning, etc. Plan who will do what, by when.

b. Plan the new facilities that are needed. No matter how small the organization or how poor its people, by faith, make plans to meet this need with the help of the Lord.

c. If the facilities do not meet standards of the local government where they are located, make the necessary repairs to come up to the required standards.

d. Repair or replace broken equipment.

e. Begin to save for new equipment that will help the organization better achieve its purposes.

f. Where applicable, purchase adequate insurance for buildings and equipment to cover replacement costs if they are damaged.

g. Where applicable, institute changes to cut the cost of utilities (heating, cooling, electricity).

PLANNING:

1. If there are no written plans, meet with key leaders of the organization to make plans.

2. If there are written plans but they are not effective, meet with key leaders to make new plans.

3. Plans should...

   -Relate to God's purposes.
   -Relate to the purpose of the organization.
   -State what is to be done.
      ...By whom.
      ...By when.
      ...If applicable, include the cost.
   -Be written so they can be evaluated to determine whether or not they were accomplished.
   -Include ways to improve weaknesses of the organization.
   -Include ways to use strengths of the organization.
   -Include steps to improve the resources (personnel, finances, facilities).
   -Include steps to improve programs of the ministry.

PROGRAMS:

1. Study the current programs of the ministry which do not contribute to the unique purpose of the organization. How can they be reorganized to contribute to the purpose? Should they be eliminated?

2. Study the current programs which do not contribute to God's purpose of spreading the Gospel to all nations. Can they be reorganized to contribute to His purposes? If not, eliminate these programs.

3. Make plans to reorganize any current program which needs reorganization. Make a list of what needs to be done by whom and by when.

4. Make plans for starting new programs that will help the organization better achieve its purpose in God's plan.

5. Eliminate any programs in which the organization is engaged which are contrary to Scriptural principles.

6. If there are no programs contributing to the spiritual growth of members of this organization, start one. If an organization is to achieve its purposes its members must grow spiritually. Plan times of prayer, worship, Bible study, and opportunities for ministering one to another.

MISCELLANEOUS ITEMS:

1. If there is no Doctrinal Statement, meet with leaders and prepare one in writing. People need to know what the organization believes and the Biblical basis for such beliefs. The Harvestime International Institute course, *"Management By Objectives,"* includes a sample Doctrinal Statement.

2. What other miscellaneous items did you identify in your analysis which need attention or correction? List these and make a plan to deal with each.

3. Make a plan to cooperate with other Christian organizations in your area for the purposes of spreading the Gospel. How can you eliminate any present problems and work together for God's Kingdom?

4. If you are a member of the organization which you analyzed, how will you use your spiritual gifts to contribute towards the fulfillment of God's purposes for this organization?

AN ORGANIZATIONAL CHECKLIST

There are several questions an organization can ask about itself in determining whether or not it is ready to minister to a particular people. Here are some questions to consider:

1. Do we have enough information about this people? Do we really understand them?

2. Have we adequate information at hand? Have we done adequate research?

3. Are we already working in this country?

4. If we are not working in this country, do we understand the complexities of working within this particular national boundary?

5. Have we ever worked with a similar people, a similar culture? What have we learned from this? Were we effective? If we were not effective, why do we think we will be effective now?

6. Do we have people available, or will we have to train them?

7. Do we have adequate mature leadership to move into the field?

8. What would be the size of the group that we think will be needed to reach this particular people?

9. How long do we think it will take before a church is planted?

10. How long do we think it will take before that church has the ability to evangelize the people within its own group?

11. Are we able to expand the present organization to carry the additional responsibility? Should we move from the place where we are now working? Is this an opportunity for withdrawal?

12. Can we expect support from those who give us financial support?

13. Will we be able to communicate what we are doing in such a way as to build the necessary prayer support?

14. Will our present structure fit this new need or will we have to change our structure? Are we able to change?

15. Will we be able to work with others who are working in this country or among this people? Will we be welcomed by the local church that may be there or others working among this people?

16. Is there another organization that can do just as good a job as we could? Why are we the ones? Have we really sought the mind of the Spirit?

17. Has God specifically called and burdened us for this people?

SELF-TEST

1. Write the Key Verse from memory.

____

____

2. What are the purposes for analyzing a Christian organization?

____

____

3. What does an analysis of an organization include?

____

____

4. As part of your study of this lesson, did you actually complete the analysis of a Christian organization?

____

5. Have you made a plan to use what you learned to help the organization spread the Gospel to the nations of the world?

____

(Answers to tests are provided at the conclusion of the final chapter in this manual.)

FOR FURTHER STUDY

Study Revelation 1:10-3:22. These chapters presents God's analysis of seven Christian organizations which were churches located throughout Asia. Complete the following chart:

Church Location	Strengths	Weaknesses	God's Plan For Improvement
Ephesus			
Smyrna			
Pergamos			
Thyatira			
Sardis			
Philadelphia			
Laodicea			

CHAPTER ELEVEN

A STRATEGY FOR PLANNING

OBJECTIVES:

Upon completion of this chapter you will be able to:

- Write the Key Verse from memory.
- Identify two natural examples which illustrate how believers are to influence their environments spiritually.
- Define the word "purpose."
- Define the word "plan."
- Define the word "objectives."
- List five steps of effective planning.

KEY VERSE:

> **Ye are the light of the world. A city that is set on an hill cannot be hid. (Matthew 5:14)**

INTRODUCTION

The lessons in this course have guided you through various types of environmental analysis. You have learned how to analyze your personal environment and that of people groups, areas, nations, regions, and organizations. These lessons emphasized that analyzing the environment is not enough in itself. You must use the results of your analysis for the purpose of spreading the Gospel.

In the natural world, you can study a field of ripening grain to determine its type and the best methods of harvest. But unless you apply what you learn...unless you actually reap the harvest...the precious grain will rot in the fields. The same is true in the spiritual world.

AFFECTING THE ENVIRONMENT

As believers, you are to affect the sinful environment of the world in which you live by being as light and salt. Jesus said He was the light of the world:

> **I am come as a light into the world, that whosoever believeth on me, should not abide in darkness. (John 12:46)**

What an impact the "light" of Jesus had on the world. His "light" showed the way from life in the darkness of sin to the glorious light of salvation:

> **For God, who commanded the light to shine out of darkness, hath shined in our hearts, to give the light of the knowledge of the glory of God in the face of Jesus Christ. (II Corinthians 4:6)**

Jesus said believers were also to serve as lights of the world:

> **Ye are the light of the world. A city that is set on an hill cannot be hid.**
>
> **Neither do men light a candle, and put it under a bushel, but on a candlestick; and it giveth light unto all that are in the house.**
>
> **Let your light so shine before men, that they may see your good works, and glorify your Father which is in Heaven. (Matthew 5:14-16)**

In the natural world, a light enables people to see in the darkness. It guides them along the road as they travel. It enables them to accomplish their work. People read, learn, and study by light.

In the spiritual world, believers are to serve purposes similar to that of natural light. We are to shine in the sinful environment around us making it possible for people who are lost in the darkness of sin to see their way "home" to salvation in Jesus Christ. Your spiritual light should enable believers around you to accomplish their ministry. People should be challenged to study and learn of God because of the glow of your spiritual lights.

Jesus also said a believer was to be like salt:

> **Ye are the salt of the earth; but if the salt have lost his savor, wherewith shall it be salted? It is thenceforth good for nothing, but to be cast out, and to be trodden underfoot of men. (Matthew 5:13)**

In the natural world, salt is used to provide flavor to food and to preserve it from decay. It also cleanses infection from wounds. In the spiritual world, you are to serve purposes similar to that of natural salt. You are to flavor, preserve, and cleanse the sinful environment around you. If you do not affect your environment, you are like salt that has lost its flavor. You serve no purpose.

By using the examples of light and salt, Jesus made it clear that believers were to be a positive influence in the world. We influence the environment to draw all men to the light of the Gospel.

The Bible is a book filled with stories of men and women of God who influenced their environment like light and salt.

PLANNING

A plan is a method or way of doing something. It is a strategy for carrying out actions.

Planning for evangelism under the guidance of the Holy Spirit is a powerful tool. It changes visions to reality, desire to demonstration. It provides opportunity for God to guide you into more effective evangelization. When you plan, you prepare to do a work for God. Planning helps you determine what needs to be done and how to do it in order to fulfill His purposes.

A STRATEGY FOR PLANNING

God has a strategy to reach each people group in the world. He has a strategy for each area, nation, and region. It is our responsibility to discover and implement that strategy.
Here are some steps for planning which will help you accomplish God's purposes. Each of these are covered in detail in the Harvestime International Institute course entitled *"Management By Objectives."*

STEP ONE: Prepare.

The first step in achieving any purpose is to properly prepare by prayer and seeking God's will.

Ask God to reveal His strategy for reaching a people group, area, nation, or region with the Gospel:

> **The preparations of the heart in man, and the answer of the tongue, is from the Lord. (Proverbs 16:1)**

You also prepare by analyzing the environment. Your purpose and plan will develop from this analysis. You must know the people, area, nation, and region in which you plan to minister. Use the skills you have developed in this course to analyze the environment.

STEP TWO: Determine Your Purpose.

Determine your purpose in God's plan for these people. A purpose is something you want to achieve. It is an end result, a goal for which you labor. Review the results of your environmental analysis. Ask God to reveal your specific purpose within His general plan to reach these people with the Gospel.

God's general purpose is that all people become one in Jesus through accepting the Gospel:

> **Having made known unto us the mystery of His will, according to His good pleasure which He hath purposed in Himself;**

That in the dispensation of the fulness of times He might gather together in one all things in Christ, both which are in Heaven, and which are on earth; even in Him. (Ephesians 1:9-10)

It is God's plan to use the Church to accomplish His purpose:

To the intent that now unto the principalities and powers in Heavenly places might be known by the Church the manifold wisdom of God,

According to the eternal purpose which He purposed in Christ Jesus our Lord. (Ephesians 3:10-11)

The Church is composed of all true believers. If the Church is to accomplish its purpose, then each believer must identify and accomplish their individual purposes. God has a specific purpose for each believer, a way in which he can serve to spread the Gospel. When you determine your purpose, you discover your part in God's plan.

When God reveals His purpose to you, write it out. This is called a "Statement Of Purpose." It is not a statement of what you believe doctrinally. It is a statement describing the specific purpose of God for you in ministering in the environment you have analyzed.

The Statement of Purpose of Harvestime International Network is provided as an example in the Appendix of this manual. The statement tells the specific purpose of this organization in the plan of God.

STEP THREE: Plan.

When you know your part in God's purpose, you must make a plan to fulfill it. You do this by setting "objectives." Objectives are written plans which...

-Relate to the purpose of God which you want to achieve.

-State what is to be done.

-Identify who is to do it.

-Set a date by which it is to be done.

-If applicable, determine how much it will cost. (Some plans can be accomplished without specific financial cost.)

It is a good idea to write out your plan. Each objective should be written in a way that you can determine whether or not it is accomplished.

STEP FOUR: Proceed.

The Bible clearly teaches that faith without works is dead (James chapter 2). This means that prayer, faith, and even planning will not be effective without action. Planning is for the purpose of action, not for the sake of planning itself. You must work to accomplish the objectives you set. Objectives must be communicated to others who will work with you. People must be trained in the skills needed, motivated, organized and directed to accomplish the objectives.

We are God's instruments to fulfill His purposes:

> **Neither yield ye your members as instruments of unrighteousness unto sin; but yield yourselves unto God, as those that are alive from the dead, and your members as instruments of righteousness unto God. (Romans 6:13)**

As faith without works is dead, so is planning without action.

STEP FIVE: Perfect.

When Jesus sent His disciples out to minister, they returned and reported the results. Jesus evaluated their ministry efforts. Evaluation is important for successful ministry. When you evaluate you examine something carefully to determine its value.

When you evaluate objectives, you determine their value in achieving God's purposes. You study the results of your ministry to determine if you achieved the plans you made. You evaluate how you succeeded and how you failed. Then you perfect the plans. You prayerfully consider the results of your evaluation to do further planning.

Your purpose will always remain the same, but the plans for achieving that purpose will change. The environment will change and you must adjust your plans. You must also eliminate objectives that did not contribute to achieving God's purposes and set new objectives and act on them.

Your purpose will remain the same, but you will repeat the other four steps of the planning cycle over and over again as you labor to fulfill God's purpose: Prepare, plan, proceed, perfect.

Always keep in mind your purpose. A purpose is like a spiritual vision. Without purpose, you will fail:

> **Where there is no vision, the people perish. (Proverbs 29:18)**

THE STUDY OF PLANNING

This lesson has presented only a brief introduction to the subject of planning. The next course in the Harvestime International Institute series, *"Management By Objectives,"* provides detailed instruction on the subject. It teaches you to faithfully manage that which God has entrusted to your spiritual care.

You can study the harvest field from now until Jesus comes, but unless you have a plan of harvest and fulfill that plan, the fields will never be reaped.

SELF-TEST

1. Write the Key Verse from memory.

____

____

2. Define the word "purpose."

____

____

3. Define the word "plan."

____

4. What are objectives?

____

____

5. What are two natural examples given by Jesus to illustrate how believers arc to influence the environment?

____

____

6. List five steps of the strategy for planning given in this lesson.

_____________________ _____________________

_____________________ _____________________

____

(Answers to tests are provided at the conclusion of the final chapter in this manual.)

FOR FURTHER STUDY

1. Study the lives of some of the great leaders in the Bible. How did they influence their environment?

   For examples, consider the lives of Joseph, Moses, Esther, Daniel, Nehemiah, Ezra, and other prophets, judges, and kings of Old Testament times.

   Study the life of Jesus and how it influenced the environment. Stephen, Paul, and Peter are other examples to study.

2. Study further about Jesus and believers as lights of the world:

   Matthew 4:16; 5:14-16; 6:22-23
   Luke 1:79; 2:32; 8:16; 11:33; 12:35; 15:8
   Philippians 2:5
   John 1:4-9; 3:19-21; 5:35; 8:12; 9:5; 11:9-10; 12:35-46
   Romans 13:12
   Ephesians 5:8
   I Thessalonians 5:5
   I Peter 2:9
   II Peter 1:19
   I John 1:5-7; 2:8-10
   Revelation 21:23

   Ask yourself: How am I influencing my environment as light?

3. Study further about believers being like salt in the world:

   Matthew 5:13
   Mark 9:50
   Luke 14:34
   Colossians 4:6

   Ask yourself: How am I influencing my environment as salt?

4. Continue your study of planning with the Harvestime International Institute course, *"Management By Objectives,"* which is the final course in the "Organizing" study module.

APPENDIX

SOURCES OF INFORMATION

Here are some possible sources of information for conducting environmental analysis. All sources may not be available or applicable to each study, but use those which do apply. Information sources include:

-Village, clan, caste, or tribal leaders
-Government officials
-Religious leaders/missionaries/ministerial councils
-Council of churches
-Headquarters of various religious denominations
-Government department of or ministry of religion
-Educators (teachers, professors, etc.)
-City hall
-Zoning Office
-City planning department
-Building department
-Utility companies
-Chamber of commerce
-Board of trade
-School district office
-Real estate boards
-Housing project developers
-Newspaper office
-Travel or visitors bureau
-Census or population reports
-Telephone company
-Local telephone directory
-Radio and/or television offices
-Transportation companies
-Library
-Government departments of records, studies on occupation, race, religions
-Local residents
-Historical and cultural museums

SAMPLE INTERVIEW

Here are sample questions to ask various leaders. Not all will apply to each analysis, but use those questions which do. You must also add other questions that apply to the specific study you are doing. These questions are only examples for you to follow:

QUESTIONS TO ASK ALL LEADERS:

1. What is your name?

2. What is your position in this_________(village, city, tribe, caste, state, nation, area, etc.)?

3. What do you see as the major problems or needs of the people?

4. How open are the people to religious change?

5. Are there any government regulations that affect evangelistic work here?

6. Do you have information on the population, educational levels, and religion for this ________(village, city, tribe, caste, state, nation, area, etc.)?

7. What is the main language spoken?

8. What are other languages spoken by the people?

9. What is the literacy rate?

QUESTIONS FOR GOVERNMENT AND COMMUNITY LEADERS:

Refer to the questions in the sections on geographical, cultural, social, and external influences in your analysis. They can help you with many of these questions.

QUESTIONS FOR CHRISTIAN LEADERS:

1. How long have you ministered here?

2. How have you succeeded in ministry?

3. How have you failed or where do you think you need to improve?

4. Are there people groups you are not reaching? If so, who, where, language they speak?

5. Is there an adequate number of true Biblical churches or is there a need for more?

6. What type of ministry are you doing here?

7. With what denomination are you affiliated?

8. Is there a need for Christian literature?

9. What type of Christian training institutes exist here?

10. Use the "spiritual" analysis section of each analysis in this course. Ask them to help you complete this portion of the analysis.

11. Explain your own spiritual purposes in this area and ask how you can cooperate with their mission.

12. Ask their opinion: Is this a time for spiritual sowing or reaping in this area?

13. What is the attitude of the people toward your ministry here?

SAMPLE STATEMENT OF PURPOSE

The importance of a statement of purpose for ministry was discussed in this course. Here is a sample statement of purpose for Harvestime International Network:

It was upon the spiritual harvest fields of the world that Jesus Christ constantly focused the attention of His disciples:

> **Say not ye, There are yet four months, and then cometh harvest? Behold, I say unto you, Lift up your eyes, and look on the fields; for they are white already to harvest. (John 4:35)**

The challenge given by our Lord is for laborers, men and women who know how to reap the spiritual harvest fields of the world for the Kingdom of God. It is to this purpose that Harvestime International Network is dedicated, to recruit, train, motivate, and mobilize a network of international harvesters capable of:

1. **Intercession for international spiritual harvest:**

    The harvest truly is plenteous, but the laborers are few; Pray ye therefore the Lord of the harvest that he will send forth laborers into his harvest. (Matthew 9:37-38)

2. **Articulation of the principles of spiritual harvest:**

    And the things that thou hast heard of me among many witnesses, the same commit thou to faithful men who shall be able to teach others also. (II Timothy 2:2)

3. **Demonstration of the principles of spiritual harvest:**

    And my speech and my preaching was not with enticing words of man's wisdom, but in demonstration of the Spirit and of power; that your faith should not stand in the wisdom of men, but in the power of God. (I Corinthians 2:4-5)

4. **Communication of the urgency of the mandate for worldwide spiritual harvest:**
    The harvest is past, the summer is ended, and we are not saved. (Jeremiah 8:20)

5. **Mobilization of members of the body of Christ to reap their appointed fields in worldwide, end-time harvest:**

    ...He reserveth unto us the appointed weeks of harvest. (Jeremiah 5:24)

ANSWERS TO SELF-TEST

CHAPTER ONE:

1. And God saw that the wickedness of man was great in the earth, and that every imagination of the thoughts of his heart was only evil continually. (Genesis 6:5)

2. The "environment" includes the physical, social, cultural, and spiritual factors which surround a person. It includes all parts of the society in which a person lives, works, and ministers.

3. The first environment of man was the Garden of Eden. It was a beautiful and perfect environment.

4. The present world environment is one of sinful man in a sinful world where Satan is still active.

5. Sin caused the change in the environment.

6. The three enemies of the world, the flesh, and the devil are in operation.

7. God sent Jesus to die for the sins of all mankind. Whoever believes on Jesus and repents from sin will be saved from spiritual death (eternal separation from God) and the eternal effects of physical death. God also has plans to destroy this present evil world and create a new world.

8. Ministry is done in the environment of a world which is basically evil. But there are many differing environments in this world. Just as all men are affected by the evil environment of the world in general, they are also affected by the specific environment in which they live. When you understand their specific environment you can be more effective in presenting the Gospel message.

CHAPTER TWO:

1. For whatsoever is born of God overcometh the world: and this is the victory that overcometh the world, even our faith. (I John 5:4)

2. An environmental analysis is a detailed study of a specific environment including physical, social, cultural, and spiritual factors.

3. The types of environmental analysis about which you will study in this course include the following:

-Personal	-Nation
-People groups	-Region
-Areas	-Existing Christian organization

4. See the seven reasons why environmental analysis is important listed in Chapter Two.

5. Six perils to avoid in environmental analysis:

   -Do not analyze things with the natural eye only.
   -Do not be overwhelmed by the greatness of the task.
   -Do not be influenced by negative reports.
   -Do not be fearful and unbelieving.
   -Do not gather facts just for facts sake.
   -Do not focus on IF, but rather on HOW.

CHAPTER THREE:

No test.

CHAPTER FOUR:

No test

CHAPTER FIVE:

1. Examine yourselves: whether ye be in the faith; prove your own selves... (II Corinthians 13:5)

2. A personal environmental analysis is a study of your own environment.

3. Personal environmental analysis includes an analysis of every area of your life including your relationships and your physical, educational, financial, and spiritual environments.

4. Here are the purposes of personal analysis:
   -To determine your spiritual condition.
   -To identify your spiritual weaknesses.
   -To identify your spiritual strengths.
   -To identify your spiritual gifts.
   -To discover your place of ministry in the Body of Christ.
   -To formulate a plan to fulfill your ministry.

CHAPTER SIX:

1. Neither is there any creature that is not manifest in His sight: but all things are naked and opened unto the eyes of Him with whom we have to do. (Hebrews 4:13)

2. A people group is any group of people where marriage and intimate life generally takes place only within the society. For example, a people group is a specific tribe, caste, clan, or lineage.

3. You study them socially, culturally, and spiritually. You also analyze external forces from outside their group which affect their acceptance of the Gospel.

4. A people group analysis identifies what people are to be reached with the Gospel, what they are like, who should reach them, and the methods by which they may best be reached.

5. If you did not actually do an analysis of a people group as you studied this lesson, please do so before proceeding with this study.

6. If you did not make a plan to reach the group you studied with the Gospel, please do so before proceeding with this study.

CHAPTER SEVEN:

1. Arise, walk through the land in the length of it and in the breadth of it; for I will give it unto thee. (Genesis 13:17)

2. For purposes of this study, an area is defined as a specific portion of a country determined by either geographical, linguistic, cultural, or political lines.

3. You analyze in detail a specific area within a nation. You determine the political, geographic, linguistic, and cultural factors that make it a separate area. You analyze its spiritual condition. You also analyze external forces from outside the area which might affect acceptance of the Gospel.

4. An area analysis identifies what people groups are in the area, what they have in common, what the area is like, who should reach the area, and the methods to use in reaching it with the Gospel, and to determine your role in God's plan for reaching this area.

5. If you did not actually do an analysis of an area as you studied this lesson, please do so before proceeding with this study.

6. If you did not make a plan to reach the area you studied with the Gospel, please do so before proceeding with this study.

CHAPTER EIGHT:

1. And that repentance and remission of sins should be preached in His name among all nations beginning at Jerusalem. (Luke 24:47)

2. A nation is composed of people united by a common political system. It is a territory with established national boundaries.

3. When you analyze a nation you study it in detail to determine political, geographical, linguistic, cultural, spiritual factors and external forces that might affect the spread of the Gospel throughout the nation.

4. An analysis of a nation is done to determine what people groups are in the nation, which have been reached with the Gospel, what do these groups have in common, what is the nation like, who should reach this nation with the Gospel, how should it be done, and discover your role in God's plan for this nation.

5. If you did not actually do an analysis of a nation as you studied this lesson, please do so before proceeding with this study.

6. If you did not make a plan to reach the nation you studied with the Gospel, please do so before proceeding with this study.

CHAPTER NINE:

1. And He said unto them, Go ye into all the world, and preach the gospel to every creature. (Mark 16:15)

2. For purposes of this study, a region is a part of the world made up of several nations located near each other geographically and similar linguistically, culturally, religiously, or politically.

3. When you analyze a region you study it in detail to determine political, geographical, linguistic, cultural, spiritual factors and external forces that might affect the spread of the Gospel throughout the region.

4. An analysis of a region is done to determine what nations are in a region, what these nations have in common, what the region is like, who should reach it, and how it should be reached. The purpose is to identify your role in reaching a region and assist you in formulating a plan to reach it with the Gospel.

5. If you did not actually do an analysis of a region as you studied this lesson, please do so before proceeding with this study.

6. If you did not make a plan to reach the region you studied with the Gospel, please do so before proceeding with this study.

CHAPTER TEN:

1. He that hath an ear, let him hear what the Spirit saith unto the churches. (Revelation 2:29)

2. The purposes for analyzing a Christian organization are:

   -To identify the purpose of the organization, its unique part in fulfilling the commission of the Church to spread the Gospel.

   -To determine the present effectiveness of the organization. How well is it currently fulfilling its purpose? How is it succeeding? How is it failing?

   -To formulate plans for the organization to more effectively fulfill its God-given purpose.

3. An analysis of an existing organization includes study of its purpose, plans, strengths, weaknesses, environment, resources, and programs.

4. If you did not actually do an analysis of a Christian organization as you studied this lesson, please do so before proceeding with this study.

5. If you did not make a plan to apply the results of your analysis, please do so before proceeding with this study.

CHAPTER ELEVEN:

1. Ye are the light of the world. A city that is set on an hill cannot be hid. (Matthew 5:14)

2. A purpose is what you want to achieve. It is an end result, a goal for which you labor.

3. A plan is a method or way to achieve purposes.

4. Objectives are statements of how you will achieve your purpose.

5. Light and salt.

6. -Prepare
   -Determine your purpose
   -Plan
   -Proceed
   -Perfect

www.ingramcontent.com/pod-product-compliance
Lightning Source LLC
LaVergne TN
LVHW060452061126
851857LV00016B/1796